Tropical Forest and its Environment

Tropical Forest and its Environment

K. A. Longman

J. Jeník

Formerly Senior Lecturers in the Department of Botany,
University of Ghana, Legon

Longman
London and New York

LONGMAN GROUP LIMITED,
Longman House,
Burnt Mill, Harlow, Essex, UK

Associated companies, branches and representatives throughout the world
Published in the United States of America
by Longman Inc., New York

First published 1974
Reprinted 1978, 1981

ISBN 0 582 44045 9

Library of Congress Catalog Card Number 73-856 81

Printed in Singapore by
Selector Printing Co Pte Ltd

Contents

ACKNOWLEDGEMENTS.

We are grateful to the following for permission to reproduce copyright material:

Commonwealth Agricultural Bureaux for the tables on pages 24 and 27 from *Tech. Comm.*, **51** (1960) of the 'Soil under Shifting Cultivation' by P. H. Nye and D. J. Greenland (Commonwealth Bureau of Soils); P. Cachan for his tables as appeared in *Annals of the Faculty of Science*, Dakar (**8**, 5–87, 1963); Macmillan Publishing Co, Inc, for a simplified table from *Communities and Ecosystems* by R. H. Whittaker (1970); The Malayan Forester for part of table (simplified) on 'Dry land forest formations and forest types in the Malaya Peninsula' from *Malayan Forester*, **27**:188–216, by R. G. Robbins and J. Wyatt-Smith; National Geographic Magazine for an extract from 'Malaysia's Giant Flowers and Insect-Trapping Plants', in *National Geographic Magazine of America*, Vol. **125**, No. 5, May 1964, page 684, by P. A. Zahl; Society for Experimental Biology Symposia for a table (modified) on page 479 of *Symposium of Soc. Exp. Biol.*, **23** (1969), by K. A. Longman, Table 1, 'Dormancy and Survival'; Geobotanischen Instituten, E.T.H. Stiftung, Rübel, for a classification with minor alterations from Ellenberg & Mueller-Dombois, *Berichte Geobot. Forsch. Inst. Rübel,* **37**:21–55 (1967).

Preface

The word 'Tropics' conjures up in many people's minds a picture of a dark, steamy and impenetrable jungle, abounding with snakes and poisonous plants. This is a distinctly misleading view to take, and indeed many types of forests exist, as well as savannas, grasslands and deserts, within the tropical regions of the world. However, it is certainly true that it is the rain forest that excites the greatest interest, for scientists no less than for the general public. In a world where space travel and molecular biology open up new horizons, tropical forests still present a baffling and intriguing challenge, the diversity of species and complexity of organisation at times defying both comprehension and classification.

Tropical forests seem so rich and productive that very high yields would appear to be possible when they are cut down and the land used for agriculture. This high potential has, however, proved peculiarly difficult to realise, partly due to inadequate or unsuitable techniques, often relying too much upon experience in temperate regions, but also because of a lack of knowledge and understanding by all concerned of the dynamic structure of the forests and their environments. Many of the biological studies carried out so far have been of the 'exploring', descriptive type, and this basic work must certainly be continued. But it is increasingly being realised that a quantitative and experimental approach is needed, both in order to solve the theoretical riddles posed by the tropical forest, and also to increase the low crop yields in many tropical countries.

The aim of this book is to assemble a dynamic picture of the trees and other plants of the forest, and the climatic and other conditions which surround them. This interacting system of forest and environment is viewed from the two standpoints of plant ecology and physiology, an attempt being made to synthesise the results of experiments carried out in controlled environments with field measurements and observations. Emphasis is laid upon major botanical concepts which are relevant to the whole tropical region, while a number of the detailed studies described are the authors' own, made in West Africa.

Only the briefest indication is given of the broader aspects of plant geography, of metabolic physiology and water relations, and of the role of animals in the forest. It is hoped that the book will be found useful by botanists and foresters, research workers, conservationists and visitors to the tropics. It will also be relevant for courses in plant ecology, whole-plant physiology, forestry and natural resources. If after reading it more students are encouraged to enter this fascinating field, with its unequalled opportunities for research, its purpose will have been achieved.

We would like to acknowledge our indebtedness to Professor P. W. Richards, whose classical reference book *The Tropical Rain Forest* remains the basic source of data on tropical ecology, and who gave us much useful advice. Mr J. B. Hall and Mr A. A. Enti gave invaluable taxonomic and other assistance in the forest, and the former together with Dr D. U. U. Okali also kindly read the manuscript. We would also like to thank Mr P. Ahn for his valuable suggestions concerning tropical forest soils, and Professors G. W. Lawson and D. W. Ewer for their encouragement. Our colleagues, and the students and the technical staff of the Department of Botany at Legon gave us a great deal of stimulation and assistance in the growth rooms and forest, even surviving the 24 h recordings of microclimate. We are particularly grateful to Mrs K. Jeníková who typed the manuscript, and to Mr J. S. Adomako, without whose careful handling of the growth data much of the physiological work could not have been completed. Dr M. R. Bowen, Miss P. M. Pears and Simon Longman are warmly thanked for their assistance with checking and indexing, and Dr. M. Rejmánek for his contribution to the illustrations. Frances Longman helped in a dozen ways, while a considerable number of unnecessary words have been omitted because of Julie Longman's emphasis on clarity and conciseness of language.

K.A.L.
J.J.

Chapter 1

Some common misconceptions

Forests are often regarded by those who live in or near them primarily as potential farm land. Depending on the type of agriculture practised, the natural tree vegetation is considerably altered or replaced entirely with other plants. In the tropics, substantial areas of the 'primary' or natural forests have been farmed for food crops by the methods known as shifting cultivation, in which temporary farms are made by cutting and burning (see Plates 1 and 12A). In many cases the forest is not destroyed completely, rapid regrowth of woody plants occurring from cut stumps and from seeds, so that the area reverts to 'secondary' forest, or perhaps to a patchwork cover of 'farm-bush' if it is cultivated again at frequent intervals (Plates 25B and 26).

In addition to providing a permanent source of land for cultivation, tropical forests yield a great deal of minor produce: firewood and charcoal, poles and thatching, fodder for animals, 'bush-meat' and palm wine, fruits and medicine. The forest is only slightly affected by the removal of small quantities of these items, and even when it was used for growing under-storey crops such as cocoa or coffee the structure of the forest was not wholly disrupted. On the other hand, plantations of rubber or oil-palm, for example, where other tree species are generally excluded, involve very profound changes from the original forest.

Clearly the use of some of the forest land in these ways provides for many of the needs of the inhabitants of the district, and helps to some extent in developing its economy. However, it may be noticed that none of these methods of land-use takes full advantage of the huge potential of tropical forests. The sheer bulk of energy-rich material they contain, coupled with the continual increase by photosynthesis, place tropical and subtropical forests in a unique position as a world natural resource (see Table 6.1, p. 121). The striking contrast which exists between this potential richness and the very low yields actually obtained is a theme which will recur several times in the following pages. Increasing these yields is a human and economic problem, indeed it is not too much to say that it is a matter of life and death to many of the people of the tropics, and yet so many attempts have ended in failure.

One of the main reasons for this failure is precisely because the forest itself has been neglected, so that the scientific knowledge was lacking upon which development should have been based. In the tropics especially, one must understand the forest in order to use it to the fullest degree and lasting advantage, and yet everywhere the story is the same: lack of official interest in conservation, continually decreasing areas under tropical forest, more concern with exploitation of a few valuable furniture timbers than in proper methods of regeneration. Even the term 'bush', which is widely used to describe both forest and many other areas with a woody cover, implies something of secondary importance, perhaps indeed a place where the rapid regrowth of trees is a nuisance, cluttering up ground intended for farming or mining, buildings or roads.

If the local inhabitants are too inclined to take the tropical forests for granted, the opposite mistake has often been made by visitors from temperate regions, who are struck at once by the great contrasts with the woodlands they have been used to. So many different kinds of trees of all sizes seem to be growing together with woody climbers, epiphytes and shrubs to form a tangled mass of luxuriant vegetation. As well as this apparent lack of order, there can be unfamiliar animals, aerial root systems, and rotting logs and whole trees, which together give a quite different impression from a temperate forest. As a result, innumerable travel books and articles have been written about the teeming life of the dark, equatorial jungle, with its climate that is supposed to be always hot and wet.

Unfortunately the picture which these authors have painted, which has been copied into geography books and taught to children all over the world, has been for the most part very misleading. As Richards (1952) has written, 'tropical vegetation has a fatal tendency to produce rhetorical exuberance in those who describe it. Few writers on the rain forest seem able to resist the temptation of the "purple passage", and in the rush of superlatives they are apt to describe things they never saw or to misrepresent what was really there.' The following extract is by no means an exception (Zahl, 1964). 'We were in a forest zone so eerie it seemed bewitched. Under and around towering trees writhed lianas as thick as a man's leg, ever straining upward in search of sunlight. Other dank vegetation seemed to clutch at us like enchanted trees in a horror film. This is the homeland of gibbons, orang-utans (the legendary "men of the forest"), wild pigs, deer and perhaps even rhinoceros. It is a haunt of cobras, and other deadly serpents too.'

One of the reasons that 'travellers' tales' are often misleading is that the writers are merely observing the forest from the outside, while passing along a road, railway, logging track or river. Such margins are frequently an impenetrable tangle of vegetation (see Plate 6); but once inside the undisturbed forest it is generally possible to walk about relatively freely (Plate 4), and it is seldom a question of continuous hacking of a pathway.

Moreover, the picture of the tropical forest as being permanently very hot needs a closer look, although it may appear a perfectly reasonable statement to anyone who is exerting himself. The human body is a bad thermometer, however, and the feeling of discomfort is due mainly to the high humidity and low wind speed near the forest floor, while it is often hotter outside. Nor is it correct to imagine that in every tropical forest water is continuously dripping from the leaves into a saturated soil. One may in fact often walk dry-shod shortly after it has stopped raining, except in wet places in the bottoms of the valleys. Similarly, to class all tropical forests as 'dark' is to ignore the variations due to species, occurrence of natural gaps, and other factors which combine to govern the intricate pattern of light and shade.

Indeed, the very idea of an unvarying climate is a false one, for there are always diurnal fluctuations, and also spatial differences between the upper parts of the canopy, the forest floor and the soil beneath, such that the leaves, trunk and roots of a particular tree experience quite different environments. Seasonal changes in climate may be slight or irregular in some tropical countries, but in many others they are considerable and predictable. For example, the tropical forest in the hills near Freetown, Sierra Leone, less than 8½° from the Equator, is subject to five months of intense dry season, with a monthly average of only 25 mm* rainfall. During this period a dry, dusty 'Harmattan' wind from the Sahara desert may blow intermittently, drastically reducing the relative humidity for a few days or weeks, and also altering the temperature and light intensity. In the rainy season, however, more than 5 000 mm of rain may fall, and tropical thunderstorms often cause disturbance to the forest. Because of the cloudy conditions, the light intensity and temperature are distinctly lower, while the day-length is about an hour longer than it is at the height of the dry season.

It is commonly supposed that tropical trees grow continuously throughout the year, but even where the climate is much less variable than in the example just quoted, rather few do so after the seedling stage. In some species in fact, the expansion of new leaves and the elongation of stems is confined to a period of a few weeks, and the shoots are dormant for the remainder of the year. Growth of the

* See page 124 for conversion diagrams to English equivalents of all metric data.

cambium may sometimes be intermittent, and this has also been reported of root growth. In addition, processes such as leaf-fall, flowering and seed germination generally show some periodicity rather than occurring at all times.

It seems indeed as though the twin myths of unchanging environment and continuous growth and reproduction have grown up together. As with most myths there is some truth in them, but they are deceptive if accepted without question. Fluctuations in climate have been judged on temperate region standards to be too small to be of importance physiologically, although neither the plant responses nor in some cases even the climate had been properly studied. At the same time, such periodicity in growth and reproduction as could not be dismissed under the heading of 'variation' has frequently been ascribed to internal or inherent rhythms, because the environment was supposed to be constant.

Sufficient has been said to make it clear that many preconceived ideas must be abandoned, both by visitors to the tropics and by those who live and work there, if the immense natural resources of the world's tropical forests are to be understood and wisely used. A thorough investigation of the complex biology of the natural forest is a formidable undertaking; not perhaps more difficult than landing men upon the moon, but certainly likely to produce more valuable results, both to theoretical science and in practical terms for the economies of the tropical countries.

Chapter 2

Forest and environment interacting

If a villager speaks of 'the forest', he generally has in mind the whole community of trees, herbs and animals, together with the dim light, moist air, equable temperature and rich humus at the soil surface. Actually, he is using this word in its most complex sense, which integrates all the forest organisms and their environment. This is very near to the advanced standpoint of modern ecology and physiology which use the term 'the forest ecosystem' in the same connection.

Alternatively, 'the forest' can describe just the tree stand, as is usual amongst European foresters. Obviously trees are the dominant features of forest ecosystems, and they are also of major economic interest, pushing the remaining organisms into the background. Strictly speaking, however, the trees represent only a part or 'synusia' of the forest community or forest 'biocoenosis'. Therefore, the term 'the forest' will be used in the following chapters to denote the whole complex of woody and herbaceous life-forms together with the accompanying animals and micro-organisms, but not the non-living part of the forest ecosystem.

The term 'environment' is also used in different senses. It may be taken broadly to cover the abiotic matter and energy surrounding and underlying a particular forest site; it is then approximately equivalent to the concept of the habitat or 'biotope'. It may also be used in the narrow sense of smaller micro-environments within the space occupied by the forest community.

Another point needing clarification is the structural range covered by the concept 'the forest'. Among foresters in tropical regions, all kinds of woody vegetation are included in this term, whether they have a continuous or a discontinuous canopy. Many 'Forest Reserves' in Tropical Africa, for example, contain only savanna-like stands with a dominant undergrowth of tall grasses. These can be termed 'open forest' (in French – *la forêt claire*), in distinction to 'closed forest' (*la forêt dense*). Since savanna vegetation has rather different ecology, it is appropriate to restrict the term 'the forest' only to closed stands, and to exclude vegetation dominated by grasses in the field-layer.

Finally, using the phrase 'the tropical forest', we are considering all the forested part of the intertropical zone, between the tropics of Cancer and Capricorn. Within this geographical area, of course, a wide range of forest types can be distinguished, such as mangrove, swampy, mesic and dry woodlands, of both evergreen and semi-deciduous habit. Particular attention will be directed in the following pages to the mesic rain forests, which have been divided by Ellenberg and Mueller-Dumbois (1967) into (*a*) tropical ombrophilous forests, and (*b*) tropical evergreen seasonal forests (see section 4.5).

2.1 The forest

Present-day ecology tends to be concerned with the complex nature and close inter-relationships among all biota (plants + animals + micro-organisms) living within one site. A mere list of species comprising the flora and fauna of the area is no longer the ultimate aim of field research. Concerted attempts are now made to understand the growth, reproduction and establishment of the organisms involved. Patterns shown by populations and the dynamics of the entire community are studied, with stress laid on relationships between the living world and the abiotic environment.

The tropical forest is an outstanding example of a complex biocoenosis (Plates 2, 3 and 4). Nowhere in the world is the diversity and the interdependence of plants and animals more obvious: shade-bearing plants germinating under the canopy of trees; lianes reaching the upper layers by utilising trees as scaffolding; stranglers surrounding big trunks; epiphyllous liverworts overgrowing leaf blades; lichens covering the bark; orchids growing in the crown-humus; fungi and bacteria decomposing wood and litter; ants feeding from floral and extrafloral glands; insects pollinating flowers; birds disseminating seeds; rodents feeding on fruits; herbivores grazing on seedlings; leopards preying on smaller mammals; and so on.

By no means is the tropical forest a mere collection of, or refuge for individual organisms, or an accidental mixture of populations. It is a dynamic system of a high order of organisation, in which the morphological, physiological and ecological features of individual members are linked together, creating forms and functions unknown outside the forest. Within the limits of inheritance, both plants and animals are modified in the forest community and forest environment, while conversely the community and indeed the habitat itself are altered by the presence of even a single big tree. Thus the varieties of the silk cotton tree (*Ceiba pentandra*) common all over the humid tropics, provide particular ranges of microhabitats for other organisms when growing within the closed forest; the branch habit here is somewhat different from that of solitary broad-crowned trees, such as the famous specimen at Freetown in Sierra Leone, under which slaves were liberated (Plate 19B).

Two important factors play a decisive role in the study and comprehension of a tropical forest: its size and the time-scale of its development. When examining a grassland one can obtain a representative sample by observing plots a few decimetres high and several metres square. An adequate sample plot in a rain forest would measure approximately 40 m in height and 4 ha in area. One cannot hope to grasp the entire structure of the tropical forest from one point on its floor, and these problems of size and perspective make all ecological observation and any experimental approach very difficult and inconvenient. The other problem in forest investigations is the great life-span and age of the organisms involved, and even more the lengthy period of development of the forest community. It is impossible for a single investigator to study the dynamics of the forest or the growth of individual trees over time-intervals of 150 years or more, which is the life-span of emergent tropical trees. Necessarily therefore, experimental studies have in most cases to be limited to the juvenile stage of tree development (see section 5.7).

Altogether the complexity, size and inconvenient time-scale create special conditions for research on tropical forests. The close co-operation of many experts is necessary, and careful selection of sample plots and ingenious planning of experiments is required. In particular, problems of growth rates, seeding, natural regeneration and survival of tropical trees can best be solved by continuing collaboration.

2.2 Geographical distribution and macroclimate

Generally speaking, tropical forests occur in the broad intertropical zone around the Equator, often just called 'the Tropics', but this is not, of course, a homogeneous, hot and humid region, wholly capable of supporting forest. Due to the particular distribution of continents and oceans, and the circulation of air masses and sea currents, there is wide variation in rainfall, air humidity, temperature, wind and other significant climatological features. Such changes in macroclimate are responsible for the large-scale pattern of distribution (see Fig. 2.1), and also affect their composition and structure.

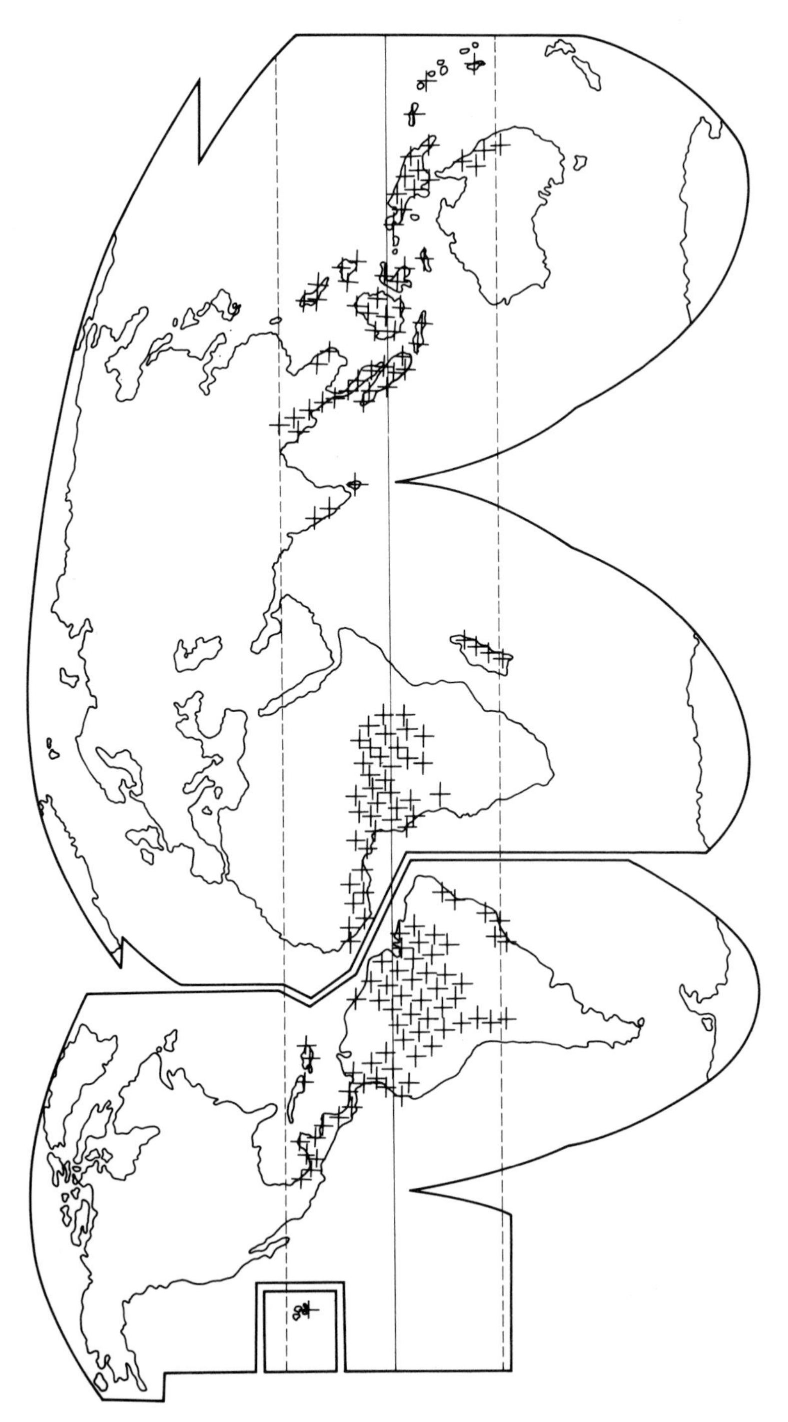

Fig. 2.1 Distribution of tropical forests. Four regions can be distinguished: the American, the African, the Indo-Malaysian and the Australian tropical-forest regions; plus the isolated Hawaiian Islands.

The climate of the tropics is governed by a potentially high level of radiant energy. High intensities of incident radiation occur because the sun is situated more or less overhead, so that the rays pass through the atmosphere by a shorter path, as compared with higher latitudes. Around midday the rays strike the surface of tropical countries roughly at right angles, and can supply a given area with more energy than in any other geographical zone. Some of this energy is absorbed by clouds, so that the drier parts of the tropics often receive more hours of sunshine than the moister regions. However, there is no winter period with greatly reduced insolation, though there are minor fluctuations as the sun 'moves' to the north or south, which can be important ecologically.

Regional differences in macroclimate are caused by exposure to trade winds, proximity and temperature of sea currents, and the extent and relief of land-masses. Critchfield (1966) classifies five types of tropical climate: rainy tropics, monsoon tropics, wet-and-dry tropics, tropical semi-arid and tropical arid climates. Tropical forests occur in the first two types, and can be considered as the climatic climax or 'zonal' vegetation of the rainy and monsoon tropics. The rest of the equatorial zone, as can be seen from an atlas, consists of savanna vegetation, or is semi-desert or desert, depending on various anomalies in the distribution of rainfall.

The principal regions with a rainy tropical climate, i.e. the Amazon Basin, windward coasts of Central America, Congo Basin, eastern coast of Madagascar, and much of tropical South-East Asia, have rainfall totals surpassing 2 000 or 3 000 mm, which are distributed more or less equally over the year. Much of this area has the potential to be covered by tropical ombrophilous forest, i.e. genuine rain forest with almost all the trees evergreen. The regions with a monsoon tropical climate – the western coasts of India and Burma, parts of extreme South-East Asia, the coastlands of West Africa, northern coast of South America, small parts of north-eastern Australia, and some of the Pacific Islands – may not differ greatly in total annual rainfall, but the year is divided into seasons of unequal precipitation and humidity. The climatic climax of these regions is tropical (or subtropical) evergreen seasonal forest, or alternatively semi-deciduous forest. In this type of vegetation some of the trees in the upper tree layer become leafless during the drier period. The severity of the dry season in tropical evergreen seasonal forest is normally much less than that found in the wet-and-dry tropics with their savanna vegetation, but extreme conditions can occur (see section 3.3).

More details about the macroclimate of tropical countries are given in a comprehensive atlas by Walter and Lieth (1960–67), from which a selection of climatic-diagrams is illustrated in Fig. 2.2. The mean annual air temperature in regions covered by tropical forests is often about 27°C. The monthly means generally lie between 24° and 28°C, so that the seasonal range is less than the diurnal fluctuations which can sometimes amount to 8°–10°C (see section 3.2). Maximum temperatures recorded in the rainy and monsoon tropics rarely exceed 38°C, thus lying far below those in the arid and semi-arid subtropical countries; indeed even in Central Europe the absolute maximum temperature may reach 40°C. Throughout the lowland tropical forest, minimum temperatures usually drop only a few degrees below 20°C. For example, the absolute maximum air temperature recorded in Santarém (Amazon Basin) is 37·2°C, the minimum 18·3°C. At the foot of high mountains and in the bottoms of valleys, temperatures may exceptionally drop below 10°C, mainly due to cold air-currents at night, for example in Malaysia or in the Cameroons. The most important fact with regard to the temperature of tropical forests is the absence of values below freezing point, which have such profound biological effects upon vegetation.

Ecologically, one of the most powerful factors controlling the pattern of tropical vegetation is the rainfall. As mentioned above, much of the area with tropical forests enjoys annual totals exceeding 2 000 mm. Favourable orography and exposure to trade winds, however, make some regions much wetter, and exceptionally up to 10 000 mm per year has been recorded (as at Debundscha, W. Cameroon). Contrary to the general impression, there is hardly any area in the world where the rainfall is distributed evenly right through the year. Even in the equatorial regions of the western and central Amazon Basin, Congo Basin and South-East Asia, rainfall generally shows some seasonal fluctuation, although this may at times be irregular. Both the degree of dryness and the duration of the 'dry' season are important factors, and become more critical for the vegetation as the total rainfall decreases. Eventually tropical forests give way to savanna woodlands and ultimately to grass savannas as the seasonal differentiation becomes more marked. Figure 2.3 shows a generalised gradient of vegetation types with respect to both rainfall amounts and rainfall seasonality.

It is well known that as tropical forests merge, on their drier and more seasonal margins, into savanna woodlands, their height decreases and they have a lower standing crop. Yet it is not commonly recognised that at the rainy end of the gradient the luxuriance of the tropical forest may also decline. In Africa, for instance, the growing-stock of ombrophilous forests, and also the diameter and height of the trees

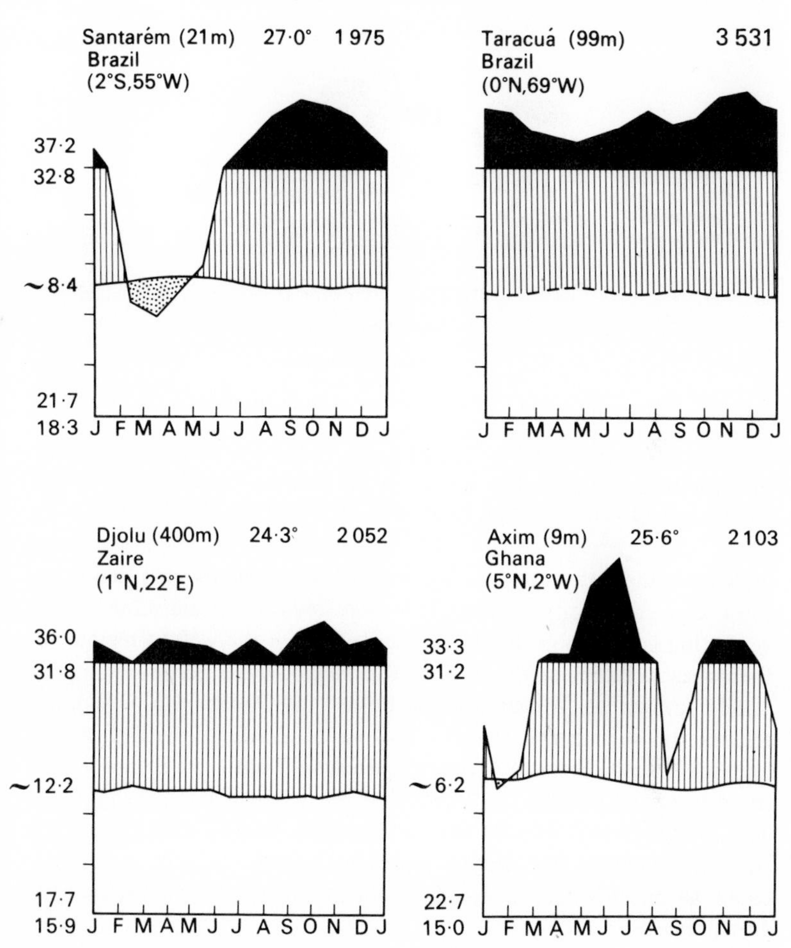

Fig. 2.2 Climatic diagrams for four characteristic but contrasting sites, situated in the rainy and monsoon tropics. The vertical intervals represent 10°C and 20 mm rainfall, except in the black layer (exceeding 100 mm per month), where the scale is reduced to 1/10. The lower curve shows the mean monthly temperatures, and the upper the mean monthly precipitation. Vertical hatching – humid period; dotted area – dry period.

Above each diagram is shown the altitude, mean annual temperature and mean annual precipitation in mm. On the left, reading from the top, are the absolute maximum temperature, the mean daily maximum temperature of the warmest month, the mean daily temperature range, the mean daily minimum temperature of the coldest month, and the absolute minimum temperature. (After Walter and Lieth, 1960–67.)

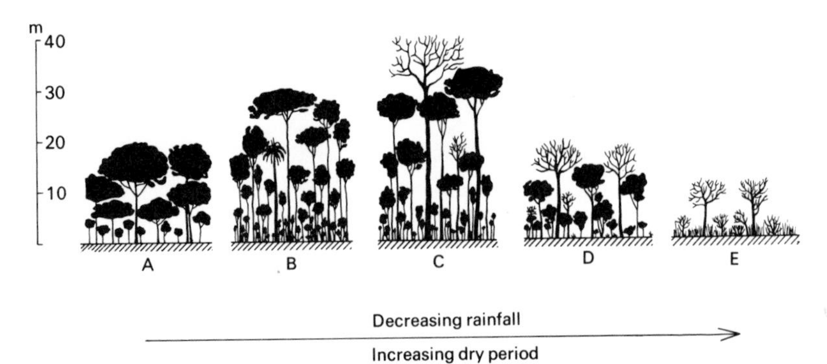

Fig. 2.3 Diagrammatic representation of one gradient or ecotone in S. America from ombrophilous forest (A and B) through evergreen seasonal forest (C) and transitional forest (D) to tree savanna (E). Note that the highest forest may not be in the wettest climate. (Vegetation structure depicted according to method of Ellenberg, 1959b.)

growing in the wettest regions, are well below the figures recorded from the evergreen seasonal forests. Still more remarkable are some of the vegetation gradients in South America, where in the wettest regions of the Amazon Basin, in the area of Rio Negro, there occur low, open stands known as Amazonian Caatingas and Amazonian Campinas (Hueck, 1966). These stands – not to be confused with the dry caatingas found on the fringes of the Amazon Basin – are composed of short evergreen trees and shrubs with a few herbs in the undergrowth. This poorer growth may be due to the highly leached soils, lacking in nutrients perhaps because of excessive rainfall. Greater cloudiness could be another factor reducing light intensity, photosynthesis and growth rates.

Any explanation and description of tropical forests would not be complete without considering the past changes of climate and landscape. Mountain building and volcanic activity were important factors influencing speciation and species diversity in the tropics. The African intertropical zone was not much disturbed by mountain building and volcanic activity during the Tertiary and early Quaternary, and this may be one of the reasons for its relatively smaller number of forest species. On the other hand, the South-East Asian islands, the Malay peninsula and tropical Australasia have been subjected, during the Tertiary, to periodic emergence and submergence of land areas, frequent volcanic eruptions, inundation by the sea and changing patterns of islands. These changes have affected not only the evolution of new biota, but also the migration and present-day distribution of plants and animals. The profound contrasts which exist between South-East Asia and tropical Australasia, as delimited by the famous Wallace's and Weber's lines, resulted from their different geological and paleoclimatological history. In a somewhat similar fashion, the vegetation of the Amazon Basin has no doubt been influenced by the neighbouring Andes range.

Considerable changes in the distribution and composition of tropical forests occurred during the Pleistocene, the beginning of which is estimated at only two million years ago. Alternating humid pluvials and semi-arid interpluvials affected the extent of closed forest in relation to savanna woodlands and grasslands. During the drier interpluvials in West Africa, for example, the area of closed forest was apparently restricted to three small refuges situated in Liberia, western Ghana and the Cameroons (see Aubréville, 1949). This restriction caused depletion of the flora and fauna, but was also responsible for the origin of new taxa within each of the isolated areas. Since during the same interpluvials the West African Forest Block was separated from the one in Central Africa, the present differences in composition between the tropical forests of these areas can be easily understood.

A number of paleogeographical factors together with differences in contemporary climate, soil and human interference indicate that the world's tropical forests can be divided into at least four major regions (see Fig. 2.1):

1. The American tropical-forest region (including parts of South America, Central America, and the Caribbean);
2. The African tropical-forest region (including the Congo Basin, the coastlands of West Africa, the uplands of East Africa and Madagascar);
3. The Indo-Malaysian tropical-forest region (including parts of India, Burma, the Malay peninsula and the South-East Asian islands); and
4. The Australasian tropical-forest region (including North-East Australia, New Guinea and the adjacent Pacific Islands).

The Hawaiian Islands may represent a smaller fifth region.

2.3 Tropical catenas

Over large areas of tropical forests, growing in the climatic types mentioned in the preceding section, the prevailing soils are ferralitic (in French – *Sols Ferrallitiques*). Sometimes called 'latosols' or 'laterites', these reddish soils when exposed along roadsides or on farms contribute to the characteristic coloration of the tropical landscape. Latosols represent the soil climax of the humid tropics under freely drained conditions. In their deeply weathered profiles, felspathic materials are decomposed slowly but completely, leaving behind simple clay minerals such as kaolinite. The mineral nutrient content is very limited, since weathering and leaching have removed nearly all bases, so that only free Fe and Al oxides, together with smaller amounts of the oxides of Ti, Cr and Ni along with fine quartz and the kaolinite are left to form the soil. (See section 3.4 and for more details Bunting, 1967 and Ahn, 1970.)

However, there is in the lowland tropics great variation between soils depending both on the relief and to some extent on the parent rock. Soil types with impeded, free and excessive drainage occur with accretion in the bottom and erosion at the top of the slopes. A sequence of soil profiles arranged according to topography is called a 'catena'. The lower part (alluvial complex) is represented by various types of hydromorphic soils such as gleys. These provide very different conditions for plant growth compared with the various types of ferralitic soils on the gentle slopes (colluvial complex) and tops (eluvial complex). Morison *et al.* (1948) made considerable use of the catena concept in their study of soil and vegetation on the savanna regions of East Africa, and it has similarly been used in ecological analysis of tropical forests (compare Guillaumet, 1967; Ahn, 1970; Lawson *et al.*, 1970).

Figure 2.4 summarises three examples of catenas from African tropical forests. The first case (A) is on slightly undulating country in which the upper and the middle complex of the catena remain weakly differentiated; here the soil of the gently inclined slopes with free drainage does not differ markedly from that of the crests. Consequently, a more or less homogeneous mesic forest covers most of the countryside, only the bottoms of valleys are wet and temporarily flooded, thus creating conditions for a tropical alluvial forest. In the second case (B), the lower part of the catena is permanently waterlogged, and tropical swamp forest with predominant palms occupies the bottom, with the remainder similar to A. In the last case (C), the three complexes of the catena are most clearly differentiated: on top there is a hardened laterite crust or bauxite layer with a drier, more open forest even containing patches of grassland; on the slopes there is mesic forest, while in the bottom of the valley permanent swamp forest occurs.

Along the big rivers like the Amazon or Rio Negro in South America, vast inland flood-plains spread over hundreds of square kilometres. The boundary between the seasonally flooded and permanently swampy part on the one hand, and the mesic area on the other, is of the utmost importance for the villagers, who make clear

Fig. 2.4 Three soil and vegetation catenas from African tropical forests. A – weakly-differentiated catena: B – moderately-differentiated catena with swamp forest on the alluvial complex; C – strongly-differentiated catena with swamp forest at the bottom and drier forest on the top (see text).

distinctions between them on practical and economic grounds. Thus 'Terra firma' with mesic ombrophilous forest is clearly differentiated from 'Várzea-Forest' on the temporarily-inundated flood-plains of the so-called white rivers and 'Igapó-Forest' in the basins of the black-rivers which drain large areas of permanently-wet peat forests. Figure 2.5 illustrates catenas from the Amazon Basin.

The variation of soils is further influenced by the parent material and the rainfall totals. In Ghana, Brammer (1962) classified three main

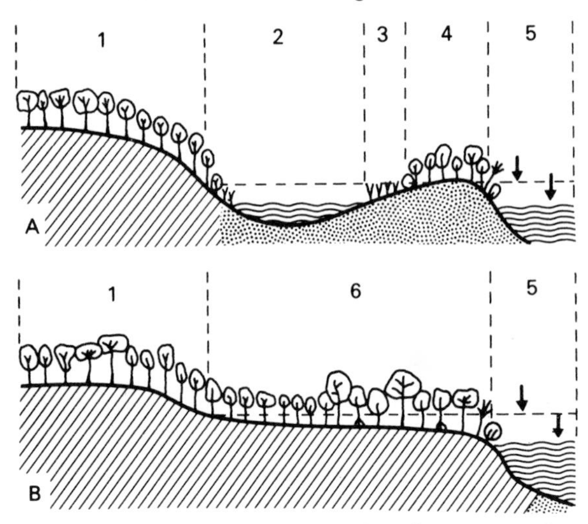

Fig. 2.5 Distribution of various kinds of tropical forest on a cross-section through a South American white river valley (A), and black river valley (B): 1 – tropical ombrophilous lowland forest; 2 – Várzea lake; 3 – swamps dominated by graminoid herbs; 4 – tropical ombrophilous alluvial forest, locally called 'Várzea-Forest'; 5 – river bed; 6 – tropical evergreen peat forest, known locally as 'Igapó-Forest'. Arrows mark normal fluctuation in water level.

soil climaxes under tropical forests. On intermediate or moderately acidic rocks in the area with less than 1 500 mm rainfall 'forest ochrosols' occur; they are red, red-brown and yellow-brown soils with a moderately acid to neutral soil reaction (pH 6–7). 'Forest oxysols' are a dominant soil climax in the regions with more than 1 500 mm annual precipitation; they develop on similar rocks to the preceding type and are much paler in colour; their reaction is very highly acid to very acid (pH 4–5). Finally, there are some 'forest rubrisols' in West Africa; they develop over basic rocks, contain montmorillonitic clay and are dark red in colour; at surface layers they are neutral (pH 7).

Close relationships between rainfall, soil and forest composition were clearly demonstrated by Ahn (1961). In the borderlands of Ghana and Ivory Coast the distribution of various types of latosols and forest associations follow very much the distribution of rainfall. Sample soil profiles showed the greater poverty of the wetter districts in exchangeable bases, total nitrogen, total phosphorus and organic matter; and this nutrient deficiency was apparently reflected in the height of the forest (25–30 m only), more simple layering (often with only two main strata), and less well developed emergent trees (see also Fig. 2.3).

Here and there within the tropical forests even typical podsols can be found, and these invariably exert a profound impact on the composition of the vegetation. In Guyana bleached sands give rise to so-called Wallaba forest; in Borneo 'Heath forest' is associated with similar soils (Richards, 1952). In some cases the extremely acid soils with pH down to 2·8 cannot support closed forest, which may be replaced by open scrub, for example the 'padang' vegetation in the lowlands of Borneo (Hardon, 1937).

Generally speaking, soils under tropical ombrophilous forest are relatively poor in nutrients, and can seldom support a continuous type of farming. Nutrient supply is better in some alluvial and swamp forests and particularly on volcanic sediments, such as can be encountered around the big Malaysian volcanoes and Mt. Kilimanjaro in East Africa. These young soils with their permanent source of nutrients from slightly-weathered minerals can support successful permanent cultivation.

2.4 Man's profound influence

When examining small-scale maps of the world's vegetation, one may get the misleading impression that vast areas in the warm and humid countries of the tropics are still under a continuous cover of closed forest. This is, however, far from true, for these maps actually indicate a kind of potential vegetation, which might be expected in the absence of man's interference.

In practice, the pattern varies a great deal. There are countries, for example some parts of South-East Asia, which are very densely populated, and here a great proportion of the 'forest' is nowadays covered with a patchwork of rice paddies and rubber plantations. Many other countries have had large areas of originally closed forest transformed into farm-bush, temporary farms and villages. Finally, there are still regions in which the primeval tropical forests stretch as a continuous carpet interrupted occasionally by big rivers: portions of the Congo and Amazon Basins, Borneo and New Guinea are good examples. In all parts of the tropics, except for South-East Asia, dense human populations in former forest appear to be of quite recent date.

Evidence from paleontology (e.g. Leakey, 1964) show that the evolution of *Homo sapiens* took place somewhere near the margins of the tropical or subtropical forests, presumably near rivers and lakes in the open savannas. Much later, man turned to the closed forest either temporarily as a hunter or food gatherer, or he was forced by inter-tribal conflicts to live there permanently. The tropical forest was therefore not a cradle but a refuge. Later, however, sophisticated societies formed within the closed tropical forest areas, for example the Mayan civilisation in America, well-organised kingdoms in Africa and highly developed communities in tropical Asia. Through his social systems man was able temporarily or permanently to overcome some of the adversities of life in the humid tropics. Recently, the density of populations has been increasing sharply, due to changes in agriculture and transport, and the introduction of foreign aid and control of parasitic diseases. These trends are causing new problems to arise in the land management of these regions.

The method of 'shifting cultivation' is still the prevailing solution which cultivators in many parts of the humid tropics have found for the problems posed by the soil. Patches of land are cleared of trees by cutting and burning, and after being used for a year or two, they are left under bush-fallow. The conversion of primary forest into farm-bush and secondary forest is proceeding very rapidly. In some countries such as Sierra Leone in West Africa, farm-bush has come to prevail almost completely (Plate 26), and the idea of a closed ombrophilous or evergreen seasonal forest being the 'potential vegetation' then becomes very hypothetical. Seeds of the majority of primary forest trees are no longer available for natural regeneration, and so the floristic composition and structure of the secondary forest may be very different from that in the primary stands.

As mentioned above and explained in one of the later sections (section 3.4), most ferralitic soils of the tropical forests are extremely poor in plant nutrients. Though these soils can carry 'luxuriant' stands of high primary production, a large proportion of the store of nutrients is contained in the living plants and in fallen logs and litter. Thus a cleared farm means a farm stripped of its major stock of nutrients; relatively few remain in the shallow humus layer and in the roots of woody plants. The removal or burning of tree trunks, limbs, branches and leaves, and the subsequent washing down of the ash is the first step to impoverishment of the site. The removal of available nutrients in the crop, and the hastened decomposition of the litter cause further degradation. Usually, the farm can be used only for one or two successive crops. About 8 to 10 years must elapse before the farm-bush or young secondary forest has restored the fertility sufficiently for further cultivation. Thus the farmer is constantly 'shifting' over the forest area, and when unexploited soil is too scarce, whole villages may migrate.

Until recently the only alternative to this cyclical type of tropical agriculture in forest areas was to make intensive use of soils situated in the lower part of catenas. Notably, rice cultivation on the flood-plains and valley bottoms has enabled South-East Asian farmers to feed very large populations in most years. Maize and cassava (manioc) are heavy-yielding crops on a great variety of better drained soils. Tree crops are often a more successful approach towards more permanent agriculture: oil-palm, rubber, cocoa, coffee, kola-nuts and citrus trees are productive cash crops which can be bartered or sold for cereals, root crops and other commodities produced in other districts.

For many ecological reasons the prospects of forestry as a method of land use in the tropical forest zone seem to be good. Tropical hardwoods appear unchallenged as veneer and furniture timbers all over the world, but so far markets have been developed for very few of the hundreds of timber species. Like raisins stolen from a cake, the desired trees are selectively extracted from the primary stands. In this way the so-called depleted forests have originated, and when not farmed subsequently they differ from the primeval forest only by the scarcity of 'mahogany' and other popular timber trees.

The building of railways, roads and channels, quarrying and mining (gold, diamonds, bauxite, etc.) are other powerful factors responsible for great changes in the tropical forests. Along the sea-shore, around big towns and in the neighbourhood of industrial centres tropical forests vanish very quickly. Unless the ecological factor 'man' places himself under reasonable control, most of them will soon disappear completely.

Chapter 3

The environment analysed

In trying to comprehend the environment of the tropical forest, man finds himself rather at a disadvantage. As mentioned earlier, it extends about him in all directions, defying him to perceive more than a small portion at once. Much of the canopy is out of sight, and the same goes for most of the rooting space. Even when using ecological and physiological instruments, gravity usually keeps him near to the ground, while the hard work required may dissuade him from digging at all deeply into the soil.

For a thorough study, the main problem is to overcome this near-the-ground viewpoint. Ideally one should observe the whole depth of the forest structure, that is 30–50 m above the ground and 2–5 m below it. Adequate observation towers and frequent soil pits are essential, though they are a rather inconvenient and often expensive way of obtaining reliable data.

One of the peculiarities of tropical forests is the spatial confusion of soil and atmospheric environments (see Fig. 3.1). In contrast to temperate forests, one cannot make a complete distinction at ground level between the rhizosphere (zone of interacting roots and soil) and phyllosphere (zone of interacting leaves and atmosphere). Soil pockets occupied by plant roots may occur at any height above the ground: in fissures in the bark, in forks between big limbs, on top of horizontal branches, and in the hollows between adjacent leaf bases. Not only a variety of epiphytes but large woody climbers and stranglers and even the supporting tree itself may sometimes put out adventitious roots into such a soil pocket.

Roots are not always strictly subterranean organs, but can grow obliquely or upwards, frequently passing undamaged through the aerial environment unless very dry conditions occur. Small pools arise in suitable hollows in leaves and tree trunks creating an aquatic environment for plants, insects, and amphibians, often specially adapted to them. The activities of animals such as ants, which often transport mineral particles and plant seeds upwards, also tend to counteract the force of gravity.

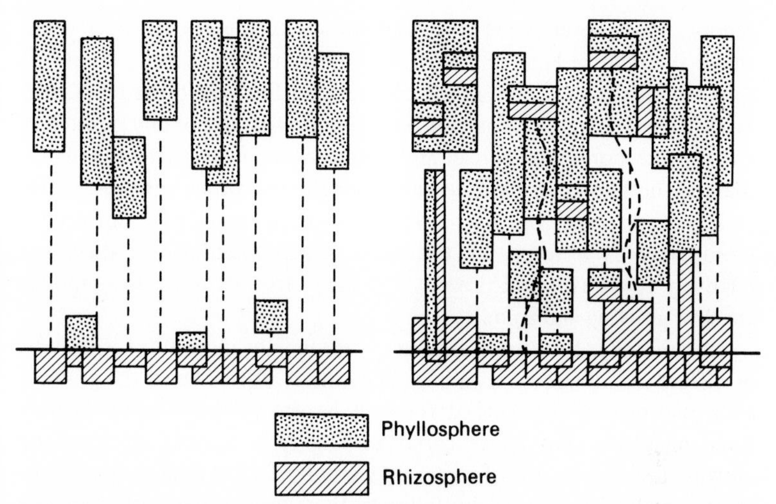

Fig. 3.1 Comparison of the spatial distribution of rhizosphere and phyllosphere in a temperate beech forest (left) and a tropical ombrophilous forest.

3.1 Light and shade

Radiant energy from the sun striking the earth's surface consists of a broad spectrum of rays. The wavelengths of the visible portion, which we call light, are very important in heating the earth's surface and supplying energy for photosynthesis. Changes in illumination, shade and darkness are important features of the tropical forest, various intensities of light and variation in its spectral composition affecting plant growth, reproduction, primary production, and thus indirectly the structure of the forest. Competition for light has been a powerful factor in the phylogeny of tropical plants, the woody climbers providing its most striking expression.

The tops of emergent trees (upper layer trees) experience the full light available, the energy they receive in a day approaching the levels striking isolated savanna trees. The uppermost twigs are often covered by heliophilous epiphytes, such as ferns, orchids and lichens (e.g. *Parmelia* spp.). A little lower down, the interior of the emergent crowns is still well illuminated, its relative light intensity decreasing according to the density of the foliage down to approximately 25 per cent of full light. This is the sphere of growth for numerous more or less light-demanding epiphytes which grow especially on the upper sides of the larger horizontal limbs, for example species of Bromeliaceae, Orchidaceae, ferns, mosses and liverworts. Below this lies the closed canopy with the crowded foliage of the middle and lower layer trees, where the relative light intensity soon drops to about 3 per cent of full light. In this layer, quite strong competition for light takes place, and various growth forms connected with this are found, notably in the leaf mosaic, the horizontal spread and arrangement of leaf blades. Finally, there is the dim layer close to the forest floor in which light intensities under 1 per cent of full light have been repeatedly measured. Presumably as a result of the very limited illumination, there are relatively few leaves and flowers, and many tree seedlings and saplings are slow-growing and spindly.

Horizontally, too, tropical forests vary in light intensity, the pattern of differently illuminated patches being most obvious near the ground. In a mature and undisturbed forest it is often possible to distinguish three phases: a *dim phase* corresponding to a 'normal' thickness of crowns, tree trunks, dwarf trees and seedlings; a *light phase* corresponding to small gaps created by fallen trees or broken limbs of emergent trees, with many seedlings germinating and sprouting in the ground; and finally, a *dark phase* corresponding to thickets of half-dead tree branches and living climbers which have fallen down from the upper layers without disrupting the canopy seriously. Only a few ground herbs and no tree seedlings can be found in the third phase.

Measurements of light intensity in the tropical forest and the interpretation of such data cause numerous difficulties. Photo-electric cells, though convenient, are selectively sensitive to different wavelengths, which are not the same as the colours absorbed by the pigments in leaves. Thus any comparison in absolute energetic units between the light intensities at tree top and forest floor, or inside and outside the forest, is rather inaccurate and physiologically unsatisfactory. Another difficulty arises from the irregular fluctuations caused by migrating sunflecks.

Several authors have tried to measure the spatial variation in relative light intensity (also called 'daylight-factor') for the tropical forest undergrowth, which is of paramount importance in natural regeneration. Sunfleck-light is a very important component which can easily be overlooked. Moreover, photosynthesis is generally greater for a given quantity of light energy when it occurs as bright 'flashes' alternating with dim light, rather than uniformly. Evans (1956) in the Nigerian rain forest and Schulz (1960) in Surinam measured and calculated the daylight-factors separately for the shade-light excluding sunflecks, for the sunfleck-light, and for the average illumination. The figures from Nigeria showed that the average daily total of sunfleck-light in the undergrowth was about 500 cal/dm^2, while the corresponding total of shade-light was less than 200 cal/dm^2. The relative light intensity for average illumination in the undergrowth was 2–5 per cent, compared with 3 per cent recorded in Surinam.

Many ecologists and physiologists have estimated relative light intensity in the undergrowth to be lower still, even a fraction of 1 per cent. For example, Bünning (1947) found the relative light intensity in the interior of an Indonesian rain forest was only 0·2–0·7 per cent; Carter (1934) reported very similar data in the *Mora*-forests of Guyana, as did Cachan (1963) in the rain forest of the southern part of Ivory Coast.

Cachan and Duval (1963) and Cachan (1963) recorded vertical gradients of illumination in an entire profile of a tropical forest. Using a steel tower they were able to measure the light intensity from the top of the emergent trees down to the forest floor (see also section 3.2). Their data show a profound screening effect of the closed canopy in the middle layer: while the illumination at 46 m at the top of the forest was 100 000 lux units, it was 25 000 lux at 33 m, still above the closed middle layer, and simultaneously only 800 lux at 1 m from the forest floor.

The light intensity near the forest floor depends partly on the structure of the canopy, and also on the angle at which the sun's rays strike the surface. The low angle of incidence in the early morning and late afternoon hours increases the path length of the light rays through the canopy (Fig. 3.2). The relative light intensity on the forest floor is considerably diminished, since the oblique rays are obstructed by more leaves, twigs, limbs and trunks, and the number of sunflecks also decreases. Leaf-movements, particularly common in leguminous trees, can increase this tendency (see section 5.4), and it can be quite difficult to read instruments correctly within the forest during the first 2½ h of the day, and again in the last 2½ h, because the light is so poor. Figure 3.3 is an example of the daily march of illumination.

Light intensity also changes seasonally for various reasons, notably the influence of cloudy or hazy periods. These incidentally also remove the sunfleck component, while the 'movement' of the sun to the north and then south may produce double peaks of light intensity (see Table 5.3, p.95). There may also be fluctuations within the forest caused by peaks of leaf-fall or of flushing. Indeed, the screening and filtering effect of the tropical forest canopy must alter the spectral composition of the non-sunfleck light reaching the lower layers. Since the intensities here are very low, such changes will be important physiologically, but unfortunately there are conflicting views on the variation in light quality which occurs. Some authors claim that there is

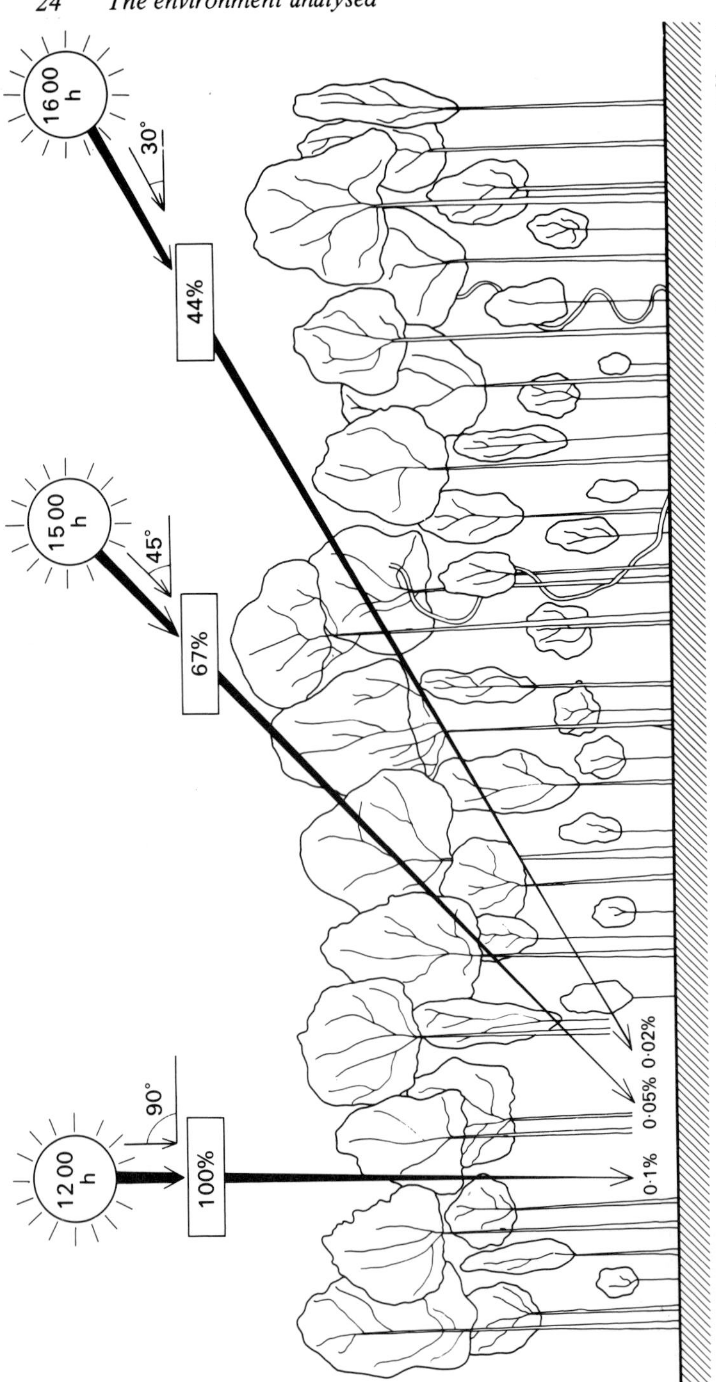

Fig. 3.2 Diagrammatic representation of the effect of tropical forest structure upon the penetration of light from different angles. Generalised picture for a tropical forest with the sun overhead at noon, showing changes at different times of day in the relative light intensity above the canopy and at the forest floor. (Jeník and Rejmánek, unpublished material.)

less red light, and others that there is more; but it seems generally agreed that there is a relatively higher proportion of far-red and infra-red (i.e. above 700 nm) transmitted through the canopy.

The day-length or photoperiod is another facet of the light environment which is sometimes ignored in the tropics. Right on the Equator it hardly changes at all, of course, but this is not to say that it has no effects upon the plants which grow there. At latitude 5°, the variation is just over half an hour, at 10° it exceeds one hour, while at

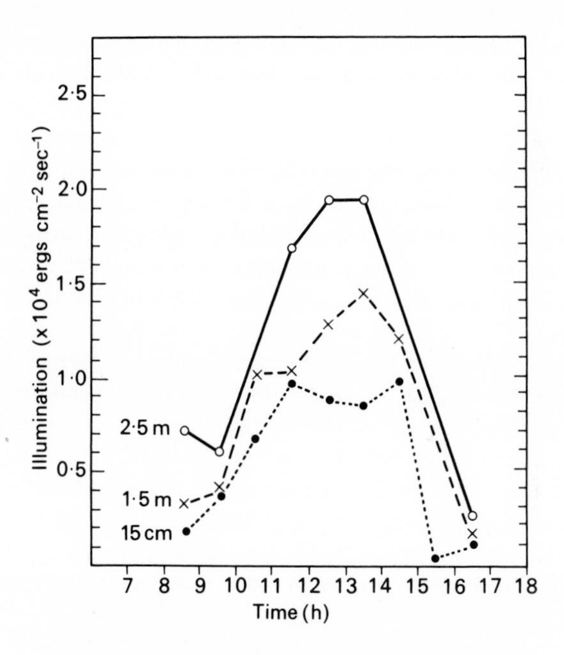

Fig. 3.3 Daily march of illumination in a rain forest in Surinam, at three levels above the ground. (After Schulz, 1960.)

17° it is two hours. As shown in Chapter 5, differences in day-length of this magnitude can result in substantially modified growth and development of tropical trees. In order to calculate the natural day-length a small allowance must be made for dawn and dusk, since plants are sensitive to quite low intensities in their photoperiodic responses. On the Equator, and at the equinoxes elsewhere in the tropics, the effective day-length is about 12¾ h for the tops of emergent trees – it would be less inside the forest or in steep terrain.

3.2 Temperature variations

The monthly average air temperature calculated for most of the meteorological stations within the tropical-forest zone is proverbially equable (see Fig. 2.2), and some writers have even concluded from this that temperature is a factor which need not be considered in the tropics (see for instance Kooper, 1927). However, behind the monthly means, diurnal and periodical fluctuations are generally hidden. Temperatures can be 5°C or more lower on days when there are storms or uniformly cloudy weather. People living in small villages or camps situated inside the tropical forest know the relatively cold nights and hot days which occur at particular periods of the year. The diurnal range of temperatures may vary periodically while the monthly averages remain the same.

Superimposed upon the diurnal and seasonal fluctuations, considerable differences in temperature occur between the various layers of the forest, and if the canopy is interrupted by natural or artificial breaks, large differences also occur horizontally. A few detailed measurements of these variations have been carried out in South America, Africa and Malaysia.

In the rain forest of the Ivory Coast, Cachan and Duval (1963) used their high tower also to study the vertical temperature gradients throughout the year. In December, the diurnal range of temperature (calculated in weekly periods) was 10·8°C at 46 m, near the top of emergent trees, but 4·4°C at 1 m height in the undergrowth. In June, the corresponding figures were only 4·0°C at 46 m, and 1·7°C at 1 m height. Data for the entire profile through the forest are presented in Table 3.1. The absolute maximum range recorded at 46 m was 14·5°C

Table 3.1. Diurnal range of air temperature (calculated for a weekly period) at various heights in the closed ombrophilous forest of the Ivory Coast.

Height above ground (m)	*December* (°C)	*June* (°C)
46	10·8	4·0
33	10·0	3·8
26	9·9	3·4
12	6·6	2·8
6	5·2	2·2
1	4·4	1·7

After Cachan and Duval (1963).

in January; the absolute minimum range at the same level was 1·5°C in June. Comparable figures at 1 m height were 5·9° and 0·7°C respectively.

Moreover, throughout the year and down the vertical profile of the forest, the daily maximum and night minimum temperatures, both very important in the ecology and physiology of trees, can vary a good deal. As averages prevail in the climatological treatment of tropical environments, relevant records of extreme values of temperatures are few, but see Table 3.2.

Table 3.2. Maximum and minimum air temperatures (calculated for a weekly period) at various heights in the ombrophilous forest of the Ivory Coast.

Height above ground	*Maximum temperature* (°C)		*Minimum temperature* (°C)	
(m)	*July*	*February*	*August*	*February*
46	25·8	32·3	18·1	23·1
33	–	31·1	–	23·1
26	24·1	30·3	18·2	23·1
12	23·8	30·0	18·7	23·2
6	–	28·5	–	23·3
1	23·8	28·2	19·2	23·4

After Cachan and Duval (1963).

The values recorded in the Ivory Coast reflect the seasonality observed even in the wettest parts of the West African forest regions. However, parallel deductions can be made concerning microclimate in tropical forests in general, for example the remarkable similarity of minimum temperatures along the whole vertical profile.

Allee (1926) has pointed out that above the tree canopy temperatures show a close resemblance with those recorded at 1·5 m height in an extensive clearing. Taking into consideration the technical difficulties of working near the tops of trees in the forest, one can appreciate the value of supplementing these studies by comparing near-the-ground measurements in the forest interior and at an opening outside the forest. Schulz (1960) collected very full data on the temperatures in different habitats of the tropical forest in Surinam. The daily maximum in the forest undergrowth usually varied between 25° and 30°C, and the minimum between 20° and 22°C. Complete curves comparing the diurnal march of temperatures in the forest interior and in a clearing are shown in Fig. 3.4, which is derived from an ecological study of the Ankasa Forest Reserve in south-west Ghana (Jeník and

Hall, unpublished material). This was done in a 'Harmattan' period, which occasionally brings rather extreme temperature conditions even to the ombrophilous forest. In other tropical regions also there can be periods of irregular temperature change. Hueck (1966) mentions the abrupt drops in temperature in the eastern part of the Amazon Basin during June, July and August, the so-called 'friagens'. At the foot of high mountains in Malaysia the temperature may exceptionally fall below 10°C even in parts of the lowland forest.

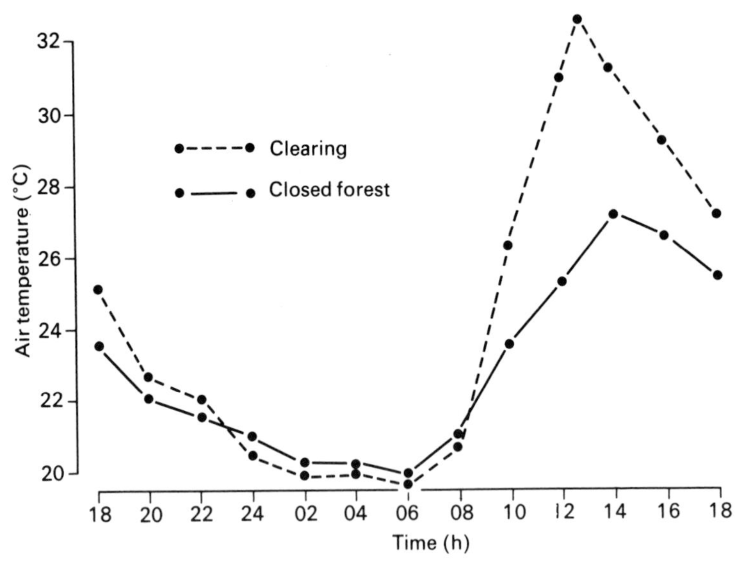

Fig. 3.4 Daily march of air temperature in an ombrophilous forest in S.W. Ghana, in a large clearing and within the closed forest. Both at 10 cm above the ground.

Soil temperatures in tropical forests in general show only minor seasonal and diurnal fluctuations. The maximum soil temperature under closed forest probably never exceeds 30°C, in contrast to open clearings where temperatures of the surface layer may temporarily surpass 50°C. Inside the forest, soil temperature seems to be controlled largely by average air temperatures, and is only slightly affected by changes in soil moisture. A comparison of the daily march of soil temperatures under closed tropical forest in a small gap and in an open clearing is given in Fig. 3.5, using data published by Schulz (1960). The same author calculated that the yearly average soil temperature was practically equal to the average air temperature in the undergrowth of the forest. At or below a depth of about 75 cm there is no diurnal fluctuation of soil

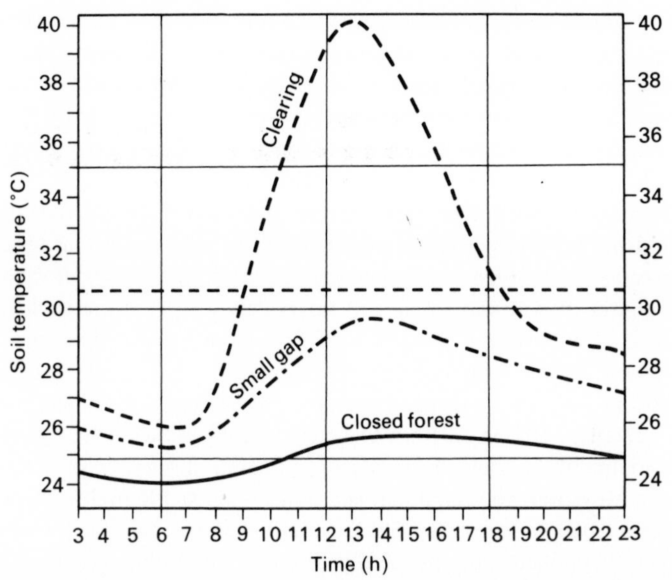

Fig. 3.5 Daily march of soil temperature in a rain forest in Surinam: in closed forest, a small gap and a large clearing. All measured at 2 cm depth, averaged over the dry season. The horizontal dotted line is for 75 cm depth in the clearing. (After Schulz, 1960.)

temperature and the seasonal range is very small indeed. Over a period of more than two years, Schulz recorded an absolute extreme range of temperature at this depth of just 1·5°C, suggesting a surprising equability of environment for root growth here. It therefore follows that a single measurement made at 75 cm depth can be sufficient to indicate the annual average soil temperature at this layer, and may even provide a satisfactory estimate of the annual average air temperature in the undergrowth of the tropical forest!

3.3 Water conditions

As explained in section 2.2, the boundaries of the area which could potentially be occupied by tropical forest can be broadly correlated with the amounts and seasonal fluctuation of rainfall. Water is frequently considered to be a key factor in the life of the tropical forest and of course plays an essential role in all its associated organisms. Many plant geographers have incorporated features connected with moisture and rain into terms defining this kind of vegetation. Schimper's expression (1898) 'der tropische Regenwald' has been widely accepted and its equivalent (e.g. the tropical rain forest, ombrophilous forest) coined in many languages; the term 'pluviisilva' also occurs in some international classifications of vegetation. In many of the subdivisions of tropical forests used by Richards (1952) and by other ecologists, words denoting moisture are very frequent: moist forest, everwet forest, monsoon forest, mist forest, cloud forest, swamp forest, etc.

Under the influence of some of these older descriptions, students in botany and forestry with little experience of tropical conditions tend to exaggerate the frequency of rains, saturation of air with water vapour, swampiness of soil and absence of water deficits. From the ecological point of view, however, most tropical forests are mesic (= mesophytic) vegetation types, situated on freely-drained soils with moisture conditions very unlike those of swamps. The variation of soil and forest along catenas (section 2.3) has to be considered in each case, though it was commonly neglected in older enumeration surveys of tropical forests. Moreover, the temporal variation in rain and humidity plays an important part in forest life. Seasonal and even diurnal changes should not be overlooked, nor the widely differing conditions on the forest floor and in the tree tops, for the 'from-the-bottom' view can be very misleading here.

Rainfall is the primary source of water in mesic tropical forests. Crowns of trees, climbers and epiphytes intercept a proportion of the rain, the actual amount varying according to the foliage and the heaviness of the fall, though 25 per cent is a good estimate of average water loss directly by evaporation from the crown (Freise, 1936). Approximately 40 per cent of rainwater runs down the limbs and trunks, and is partly absorbed by dry bark, partly evaporated in the forest interior, while part reaches the soil at the base of the trunk. Thus only about one-third of the rainfall can be expected to penetrate directly through the canopy of the forest, and consequently, the usual types of rain-gauge greatly underestimate the total rainfall.

Other sources of moisture are dew, fog and low clouds. During a cloudless night and the following morning, dew is an important phenomenon and may amount to 0·1–0·3 mm, i.e. 100–300 ml of water/m^2 (Walter, 1962). On the ground, dew can be encountered only in larger clearings, for in the closed stand the level of radiant cooling and condensation is shifted upwards to the top of the crowns. The particular shape of the crown and the form, size and position of leaves may modify the rate of condensation. In tropical Africa, for instance, abundant dew is often formed on the large, digitate leaves of the 'umbrella tree' (*Musanga cecropioides*), and this can be seen dripping down to the ground. In other cases the drops of dew may not be so large, and the leaves or leaflets may be smaller, so that the water does not reach the ground. However, its presence on the leaf surface for the first 2–3 h of the day can reduce transpiration and thus influence the daily water balance of the plant. Fog and low clouds affect the ecology of tropical forests mainly in the mountains, and along rivers and the sea shore. Ellenberg (1959a) describes tropical 'fog-oases' in the desert coastlands of Peru, where a low forest occurs at certain altitudes and the tree branches 'comb out' minute droplets from clouds drifting in from the Pacific. Similar striking effects of cloud and fog are found in the mountains of East Africa and South-East Asia.

In the life of tropical forests various aspects of atmospheric moisture have been correlated with plant growth. The absolute amount of invisible water vapour has no particular ecological significance, but an important consideration is the relative humidity, the percentage of the maximum quantity of vapour which can be held by the air under a given temperature and pressure. Numerous measurements have been made to find out how much truth there is in the idea of 'permanent humidity' in the tropical forest. All available figures suggest that there is in fact great variation in space and time of this environmental factor.

The day-time figures for relative humidity are particularly variable. High in the canopy during the mid-day period, the level may fall to 70 per cent, while close to the ground in the forest interior it is generally above 90 per cent. However, in regions affected by seasonal rainfall fluctuations, day-time figures may be encountered which suggest considerable though temporary dryness. The frequency and duration of such times of increased water stress may influence the floristic composition of the forest and the rate of shoot growth (section 5.2). For example, during the 'Harmattan' season in African forests, the relative humidity of the air near the forest floor may fall almost to 70 per cent, a value so low as to damage the 'stenohydric' plants in the undergrowth (see Fig. 3.6). At a similar time of year, Cachan and Duval (1963) even recorded 30 per cent relative humidity of the air at 46 m near the tree tops. All these figures show that plants (or plant organs) and animals alike enjoy or perhaps endure in the upper canopy an entirely different day-time microclimate from that in the undergrowth. Related to this may be the different size and perhaps shape of leaves of the same tree in various layers above the ground (see section 5.7).

In most cases there is a long nocturnal period during which relative humidity remains above 95 per cent, and frequently approaches the point of full saturation. The pertinent observations show a similar pattern during the night in all layers of the forest, including the ground

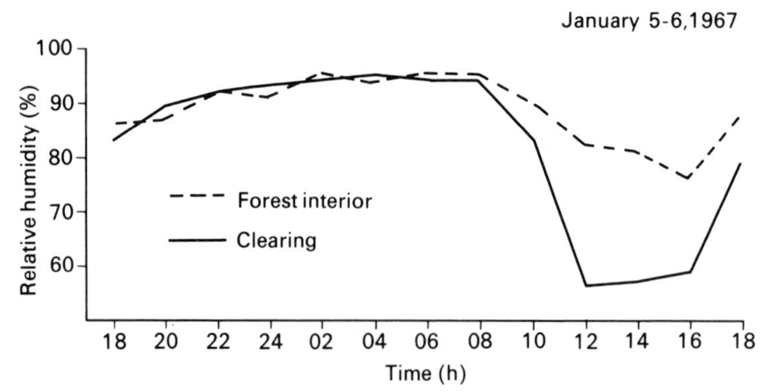

Fig. 3.6 Daily march of relative humidity in an ombrophilous forest in S.W. Ghana on an unusually dry day during a 'Harmattan' period. Both readings at 10 cm above the ground.

layer of the clearings, and it seems that the same applies even in drier seasons (see Fig. 3.6).

A shortcoming of relative humidity data is that they do not express accurately the drying power of the air, but the saturation deficit and vapour pressure deficit are better since they take account of the current temperature and barometric pressure. Perhaps the best ecological and physiological index of the aerial environment is the evaporative power of the air, because this also allows for the effects of the wind. For a comparison of different measures, see Table 3.3.

Measurements in the undergrowth with various kinds of evaporimeters generally show exceptionally low rates of evaporation, chiefly because of the near-saturation of the air and the lack of wind near the ground. Naturally, there is little evaporation at night, and in the day-time near the ground it is usually only one-quarter or less of the levels achieved in the upper layer. During the rainy season in West Africa, for example, there are many 24-h periods without any marked evaporation close to the ground, but on the other hand during the dry season rather high values may temporarily be recorded. In the upper layers the trees may sometimes be exposed to rates of evaporation which are comparable to those occurring in the savanna (Jeník and Hall, 1966).

Both the low rate of evaporation in the undergrowth and the high water stress in the canopy raise many difficult physiological problems. In the saturated atmosphere with little transpiration the usual pattern of mineral nutrient transport in the xylem may perhaps not occur. Is this a contributory reason for slow seedling growth and the stunted appearance of many pigmy trees in the undergrowth of tropical forests?

Table 3.3. Diurnal fluctuation of humidity near the ground in the tropical ombrophilous lowland forest during a dry spell of 'harmattan' weather 'Ankasa Forest Reserve, Ghana, 6–7 January 1967); the evaporative power of the air measured by Piche evaporimeter with green disc 3 cm in diameter.

	Time (h)											
	20	22	24	02	04	06	08	10	12	14	16	18
Forest interior:												
Relative humidity (%)	94	95	91	96	93	97	92	85	71	69	71	77
Saturation deficit (mb)	1·6	1·3	2·2	0·9	1·7	0·7	1·9	4·4	9·3	10·9	10·4	7·4
Dew point (°C)	20·9	20·8	19·6	19·6	19·3	19·3	19·7	20·7	19·4	20·9	20·8	20·8
Evaporative power (ml/2 h)	0·1	0·1	0·0	0·0	0·0	0·0	0·0	0·1	0·2	0·3	0·3	0·2
Forest clearing:												
Relative humidity (%)	87	91	94	94	93	95	94	70	51	45	60	75
Saturation deficit (mb)	3·4	2·4	1·4	1·4	1·7	1·3	1·4	10·4	22·1	24·9	16·0	8·0
Dew point (°C)	20·5	20·4	19·5	19·3	19·1	18·9	19·7	20·4	19·6	17·9	20·7	20·2
Evaporative power (ml/2 h)	0·1	0·0	0·0	0·0	0·0	0·0	0·3	1·2	2·2	2·1	1·2	0·4

On the other hand, can the prevailing xeromorphic structure of leaves in many tropical trees be regarded as an adaptation to moisture stress encountered during sunny days on the crown surface and at ground level in clearings?

If rainfall is the main factor controlling geographical distribution of tropical forests, soil moisture is dominant in determining local patterns of forest types. Obviously, the position within the catena affects the degree of drainage, and consequently the water content in the rhizosphere. As shown in Fig. 2.4C this includes sites with impeded drainage, free drainage and excessive drainage. Unless there are some irregularities in the soil profile, for example the presence of an indurated hard-pan, habitats of impeded drainage are limited to the very bottoms of valleys and the flood-plains of large rivers. These water-logged sites are distinct in floristic composition and overall structure of the forest: recognisable types are developed such as tropical ombrophilous alluvial forest (riparian, occasionally-flooded or seasonally-flooded), tropical ombrophilous swamp forest, tropical evergreen peat forest, etc. (see classification in section 4.5). The prevailing forests, however, are mesic types, growing on slightly impeded and particularly on freely drained soils. Even during the height of the rainy season it is possible to walk in these forests without rubber boots, the conditions usually resembling those encountered, for example, in European beech forests.

Continuous measurements of soil moisture in tropical forests have been few. Schulz (1960) confirmed that soil moisture is strongly correlated with the texture of the soil. Bronchart (1963) recorded the variation in soil moisture in the surface layers of evergreen seasonal forest at 0° 30′ S in Zaïre (Congo, Kinshasa) during a whole year (Fig. 3.7), and disclosed the surprising fact that water contents may drop below the permanent wilting percentage even in this vegetation type. In his study Bronchart also provided experimental evidence that fluctuation in soil moisture may control the seasonal flowering of the ground flora (see section 5.7).

Markedly different conditions for plant growth are found in the water-logged soils with impeded drainage. After periods of rain, these soils may be temporarily flooded, the duration of the flooding being from several hours to several weeks. Photographs showing tropical investigators walking knee-deep in water through the forest are almost certain to have been taken at a low point of the catena. The excess of soil moisture interferes with aeration, another factor of paramount ecological significance. Restricted aeration prevents numerous species of the mesic forest from colonising the alluvial complex of catenas.

Compared with the mesic types. alluvial and swamp forests have fewer constituent species, and in the latter type there is a tendency towards single-dominant stands. Palms are very successful life-forms in this environment in many parts of the world, and even in the herb layer large monocotyledonous plants predominate (Marantaceae, Zingiberaceae, grasses). Locally the forest may even give way to patches of savanna vegetation. Some dicotyledonous trees still flourish here, particularly if they have special adaptations assisting in gaseous exchange (see Fig. 4.5, p. 58).

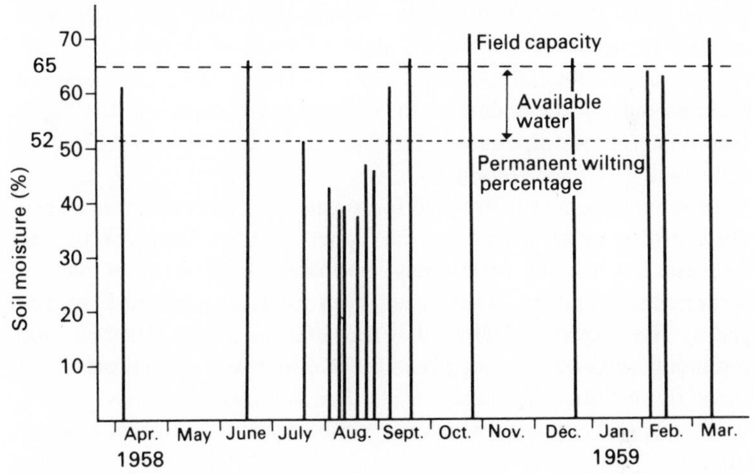

Fig. 3.7 Seasonal changes in soil moisture in a rain forest near the Equator in the Congo basin. Soil moisture content expressed as a percentage of fresh weight. Note the levels well below the permanent wilting percentage during August. Rainfall measured at the same site varied from 0 to 124·7 mm per 10-day period. (After Bronchart, 1963.)

At the other end of the moisture gradient, dry soils can develop locally, on the top of the catena or over excessively drained sands. Within these sites the growth of the forest is reduced and drought-tolerant trees and grasses may occur in numbers. In some parts of the tropics one can even find savanna-like patches deep inside ombrophilous forest. In Zaïre these and the wetter patches, called 'Esobe', are even found close to the Congo River (Germain, 1965), while in West Africa they can be found on indurated bauxite deposits capping very ancient hill-tops. However, in all these cases man's interference could have played an important role in expanding the area or numbers of these savanna patches, because they have often served as centres for settlement and cultivation of crops.

Permanent or temporary excess of moisture inside tropical forests is a factor which can be correlated with various growth forms, environmentally-induced modifications and adaptations. The shape of leaf blades is an interesting example of epharmonic convergence in tropical forest trees, woody climbers and shrubs: their caudate leaves are very frequently extended into a narrow acumen called a 'drip-tip', which is usually curved downwards. In a study in the Ankasa Forest Reserve, Ghana, it was found that more than 90 per cent of species in the forest undergrowth possessed leaves with drip-tips. In an experiment comparing intact leaves with leaves deprived of the tip, the former were absolutely dry only 20 min after rain, the latter were covered by spots of water even after 1·5 h. In conformity with the vertical environmental gradient many emergent trees may develop drip-tips only in the undergrowth, while leaves occurring in the upper canopy are obtuse or truncate, for example *Lophira alata* in Africa (see Plate 7a, Fig. 4.7 and section 4.2).

Other peculiarities of tropical forests are the 'arboreal swamps' and small 'water tanks' found in the upper canopy. They are formed amongst the closely overlapping leaf-bases, both of trees and of herbaceous epiphytes. They may even contain specialised aquatic plants, like certain species of *Utricularia* in South America, and commonly serve as breeding places for tree frogs and mosquitoes. Erect waxy flowers such as those of *Bombax buonopozense* may provide small temporary 'pools', which are visited by ants and other arboreal animals.

3.4 Availability of nutrients

The luxuriance and high primary production of tropical forests sometimes lead to the false assumption that the soil supporting such vegetation must contain a large supply of mineral nutrients, and that the same reserves can be utilised for an unlimited period by agricultural crops. However, both the accumulated experience of villagers and the yields achieved by agriculturists have shown that the fertility of most freely-drained soils under tropical forests has a short duration when used for agriculture. After two or three crops have been harvested within a short period of one or two years, yields quickly fall. This contrast between permanent luxuriance of the primary forest and quickly declining crops can be explained in terms of great changes in the amounts and availability of nutrients. (See Nye and Greenland, 1960).

Ferralitic soils are rather uniform in their clay composition and overall chemical characteristics. The clay fraction consists mainly of kaolinitic minerals with iron and aluminium oxides, and very few undecomposed minerals rich in nutrients are left within reach of plant roots. Moreover, the cation exchange capacity of these soils is very low, except for the thin layer of humus at the top. In high-rainfall regions the latosols are also highly acid, with soil reaction as low as pH 4. The C/N ratio tends to be high, with figures well over 10 being commonly found. Detailed chemical analyses have been published in many countries: see for example Owen (1951) and Ahn (1961).

Table 3.4. Nutrient elements (in kg/ha) stored in the upper 30 cm layer of the soil, together with those stored in the biomass plus litter, in a mature secondary stand within a tropical rain forest.

	N	P	K	Ca	Mg
Soil	1 830	125	820	2 520	345
Stand	4 580	12	650	2 580	370

After Nye and Greenland (1960).

The nutrient cycle in the tropical forest cannot be properly assessed without an appreciation of the nutrients stored in the vegetation. Analyses of ferralitic soils have revealed the striking fact that the quantity of some exchangeable nutrients in the topsoil may be of only the same order as the amounts contained in the biomass of the living forest. Nye and Greenland (1960) undertook the laborious task of making a large number of analyses and calculations, a few of which are reproduced in Table 3.4. It is important to realise that the nutrient cycle and also the accumulation of organic matter in the tropical forest is very different from the same processes in temperate woodland ecosystems (Fig. 3.8).

Once the primary tropical forest has been destroyed either by extensive extraction of timber or by burning and farming, the natural nutrient cycle is damaged and the humus content falls. Mineralised nutrients are quickly washed away or leached by rainfall, and in addition a considerable loss may occur when crops are harvested. Altogether the soil conditions typically deteriorate, and farming generally ceases.

The above account is, however, imcomplete when long-term experience with 'shifting cultivation' is taken into account. The bush fallow that develops after farming partly restores the fertility of the soil, though not to its previous level. Among the trees involved in the succession, many are found to be eutrophic species, demanding a rather high concentration of nutrients. The deep penetration of tree roots allowing the lifting of leached nutrients has been offered as an explanation for this point. Although, as will be seen later, tropical trees are not as shallow-rooting as it is often claimed, the general distribution of the feeding roots does not allow such an assumption, for the nutrients are generally leached to much greater depths.

It appears that the nitrogen cycle is the decisive factor here, for forest trees and their associated synusiae of plants and micro-organisms can rather readily restore a depleted supply of nitrogen. The balanced co-operation of numerous groups of micro-organisms, including nitrogen-fixers, may play an important role; and Ellenberg (1959b) also stresses the importance of mycorrhizae. Additional nutrients may be made available through the process of chelation, since many forest plants and micro-organisms appear to produce chelating agents which participatc in soil processes. In most cases experience shows that the restoration of fertility takes some time, so that it is desirable to leave the land under bush fallow for 10 years or more before farming again.

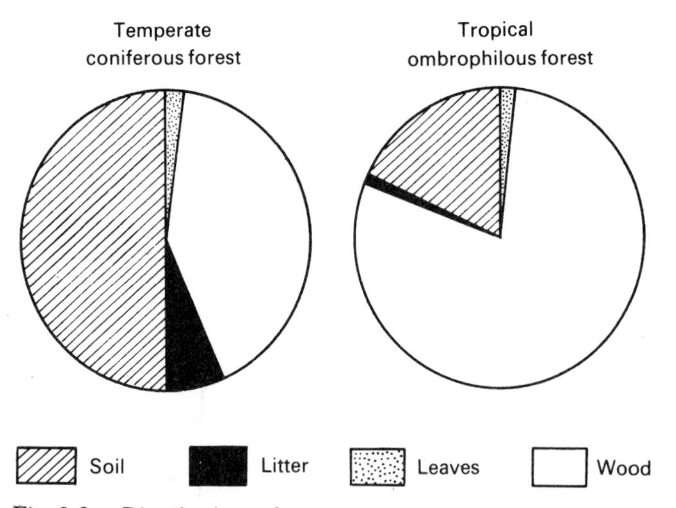

Fig. 3.8 Distribution of organic carbon in the abiotic portion (soil, litter) and biomass (wood, leaves) of tropical ombrophilous forest and temperate coniferous forest. (After Kira and Shidei, 1967.)

The different species composition of the secondary regrowth does not necessarily indicate deterioration of soil fertility. Some species are killed by fire, and do not readily colonise the abandoned farm from the surrounding forest. The availability of seeds, the conditions for their germination and the coppicing power of the species may be important. However, irreversible impoverishment of the soil profile often does take place, particularly when farming is done more frequently and rainfall is high.

The soils in evergreen seasonal forests change their characteristics during the year. For instance, nitrates, the major source of nitrogen for plants, are very soluble and may be leached down the soil profile during the rainy season, where they may not be readily available; at the same time denitrification may be further reducing the supplies. In the dry season, however, especially towards its end, increased quantities of nitrates are often recorded in the topsoil. The role of lightning in fixing atmospheric nitrogen is not fully understood but it may be important here. The consequent improvement of nitrogen supply with the early rains may influence the growth of leaves, stems and fruits, which, as discussed in Chapter 5, is frequently taking place at this time. (See Table 5.3, p. 95).

The role of nitrogen-fixing organisms in the tropical forest is still an open problem. Among the trees, for instance, leguminous species are fairly frequent, and consequently, the importance of *Rhizobium* symbiosis might be anticipated. However, in excavations of roots in West African forests, many species of Papilionaceae, Caesalpiniaceae and Mimosaceae apparently had not developed root-nodules. More important seem to be the free-living aerobic bacteria of the genera *Beijerinckia* and *Azotobacter*, the former being specific for high-rainfall regions with highly acid soils. Blue-green algae and fungi, both epiphytic and endophytic, include many nitrogen-fixers, and under anaerobic conditions *Clostridium* spp. take part in nitrogen fixing (Meiklejohn, 1962). The photosynthesis of the forest ensures a continuous supply of the energy-rich materials necessary for nitrogen fixation by the heterotrophs. Conversely, the nitrogen-fixing organisms supply some of the nitrogen necessary to maintain a balanced nutrient cycle. Fertiliser experiments in fact often suggest that within a mature tropical forest no deficiency of nitrogen can be found, though in the same site under farming this is seldom true.

Apparently related to the concentration of available nutrients in the surface layer, the roots of tropical trees often spread mainly in this part of the soil. It is nevertheless possible to exaggerate the shallow-rootedness of these trees. Excavations and semi-quantitative estimates in Ghana (Mensah and Jeník, 1968) showed that the shape of the root skeleton of *Chlorophora excelsa* does not differ markedly from that of European oaks which serve as text-book examples of deep root systems. Tap-roots and the large vertical sinkers can penetrate to a depth of at least 2 or 3 m. It is the distribution of small feeding roots which seems to be limited more strictly to the upper 5 cm layer of the topsoil (see

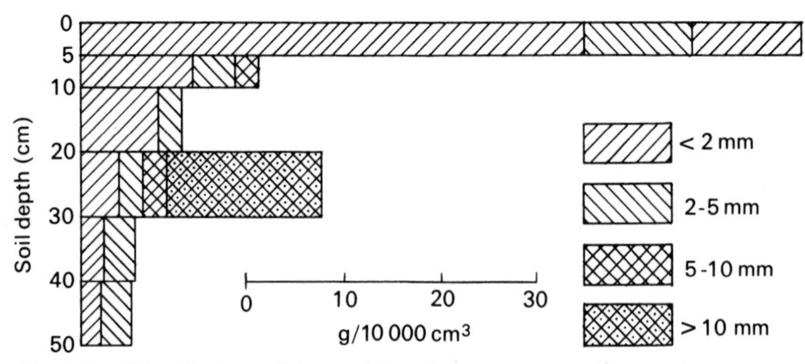

Fig. 3.9 Distribution of dry weight of tree roots at different depths in an evergreen seasonal forest in Ghana. The horizontal scale represents the amount of root material per unit volume of soil in four diameter classes.

Fig. 3.9 and Lawson *et al.*, 1970). In temperate forests, most of the small feeding roots are to be found in a rather thicker layer of about 10–20 cm depth, but in both cases there is an obvious correlation with the humus and main nutrient source.

As noted earlier, it is not unusual to find considerable quantities of crown humus in tropical forest. This consists of debris of all kinds: decaying leaves, twigs, bark, wood and fruits, mixed with mineral soil brought by ants from the ground. In the first instance, accumulation of crown humus occurs anywhere in the forest canopy where large limbs spread horizontally, tree trunks fork or rough bark covers the stem. Terminal rosettes of leaves on pigmy trees in the undergrowth may also collect crown humus. Epiphytes germinating in the initial pocket of soil later contribute substantially to the rate of accretion. This can happen simply because the epiphyte is there, or because of special adaptations which increase the amount of humus trapped, for example, the nest-fronds of some ferns (*Platycerium* spp., *Drynaria* spp.) and the brush-roots of some orchids (*Ansellia africana*, etc.). Gradually quite a thick carpet of crown humus can cover the upper surface of the bigger limbs (see Plate 5), where it is freely penetrated by the roots of the epiphytes and exceptionally also roots from the 'host'. In older trees the total weight of the burden carried may prove too great and a limb or even the whole tree breaks, bringing down the whole community. Such catastrophic occurrences open a new chapter, paving the way for regeneration of the forest (see section 4.1).

3.5 The impact of wind and fire

On a global scale, wind has an important influence on the geographical distribution of tropical forest through the effects on the direction of movement of moist and dry air masses. Moreover, the pattern of vegetation in a particular region may reveal the influence of wind, which can also affect the life of the forest community in many ways.

On the seashore, low littoral scrub merging into normal rain forest is usually associated with salt-laden winds. Similarly, the dwarfing of the forest and the presence of flag-shaped trees on the top of wind-swept mountains can be attributed to the mechanical and physiological effects of wind. Frequent cyclone tracks through tropical forests may be clearly distinguished by their structure and floristic composition. Thus Beard (1945) mentions 'hurricane forest' in the West Indies, Wyatt-Smith describes 'storm forest' in Malaya, Webb (1958) studied 'cyclone scrub' in the tropical lowland forest in North-East Australia, and Jones (1956) stresses the importance of the disturbing role of tornadoes in Nigerian forests.

The effect of frequent cyclones overrides the usual ecological factors of soil and microclimate. Instead of a stable forest climax, a disturbed community is found with a helter-skelter of trees, lianes, vines and herbs covering the damaged area. Repeated hurricanes deplete the forest of its bigger trees and pioneer species may become dominant.

Wind-speeds recorded during these violent storms may locally reach 150 km/h. The whole upper canopy can be defoliated, large limbs of emergent trees broken, and many trees uprooted. Exposed slopes and hill-tops may not maintain closed forest at all, and repeated catastrophes can result in a complete change of species. The opinion has been expressed that the disturbing role of cyclones has played an important part in the speciation of tropical plants and in the evolution of tropical vegetation (Dobzhanzky, 1950). The formation of long horizontal roots in *Entandrophragma angolense*, spreading just above the soil surface, has been related to the stresses imposed by violent storms (Jeník, 1971a).

From the point of view of the internal organisation of tropical forests, even occasional breakage of crowns and sporadic uprooting of trees can have far-reaching implications in the life of the forest community. Except where other powerful mechanical factors such as elephants are at work, only wind interferes profoundly with the forest equilibrium, changing at a stroke the conditions for regeneration, growth and reproduction in the lower layers. The clearing caused by a single wind-thrown tree may spread over half a hectare, since other nearby trees linked by climbers can be dragged down or snapped. Due to this factor, many heliophilous species are able to survive in large blocks of 'closed' forest.

Calm air usually prevails in the interior of the forest, but slow currents can usually be measured on the margins of larger gaps where circulation is induced by temperature differences between the closed forest and the open ground. It is therefore not surprising to find that, in comparison with many other types of vegetation, wind pollination and wind dispersal of fruits and seeds is less dominant (see sections 5.8 and 5.9).

The importance of fire in the inhabited tropical forests is widely accepted. In the hands of the villager, fire is an efficient tool in clearing his farm and protecting the village against encroachment by the forest. However, at no time of the year does fire spread in the closed forest, as it does in many other types of woodland, since there is insufficient dry material on the forest floor to sustain a 'crown' fire. If a substantial number of the trees is felled, burning can follow when the wood is dry enough. Thus it is not easy to visualise many natural conditions under which fire affected the tropical forest before human immigration.

Lightning and volcanic eruption are the main natural sources of fires, and they rarely coincide with patches of desiccated forest. Probably trees are occasionally set on fire by lightning in cyclonic paths, and this certainly occurs along the edges of lava streams, where large areas of forests are killed and partly burnt by hot lava and falling ash. In South-East Asia and around the large African volcanoes, the disturbing influence of fires together with the nutrient-rich soils derived from lava flows have probably contributed to the richness of the tropical flora.

In the transition zone between tropical forests and savannas fire is a much more general factor. The more pronounced periodicity of the climate and the presence of grasses create conditions for the natural spread of fire during the dry season. Some ecologists consider that the present boundaries between the closed forest and the savanna are the result of many years of fires during which the savanna expanded deeply into the forest zone, and terms such as 'savannisation' or 'derived savanna' are used in this connection. With the addition of man-made fires, is this the future of the entire forest area in the tropics?

Records and observations show that burning can permanently destroy the forest in favour of savanna or grassland only in regions situated in the transition zone, where the ecological balance between the trees and grasses can easily be upset. Walter (1962) summarises the major characteristics of both competitors with regard to their fire-resistance. Riverain forests existing amidst savanna vegetation which is annually burnt, provide evidence that closed forest properly supplied with water cannot be destroyed by fire. In densely inhabited regions, fires have reduced the transitional zone to a very narrow strip, while in some areas there is an abrupt change from forest to savanna.

Present-day systems of shifting cultivation cannot be practised without burning. Though in the long run their destructive effect on the humus and the nutrient cycle are undeniable, these methods work, and we shall probably have to accept fire as an unavoidable factor in tropical forests for a long time to come.

3.6 Some biotic factors

Plants, animals and micro-organisms participating in the tropical forest community are interconnected by numerous direct and indirect relationships. In a given area some of the species are present in very large populations, for example bacteria, algae, insects; some occur in limited numbers, such as the smaller trees, herbs, rodents; while some species may appear as isolated specimens, for example emergent trees, big mammals, snakes. Within the community various populations can be grouped into specialised synusiae which have more or less similar form and function and occupy also a definite position within the forest.

Following Richard's classification (1952), one can distinguish the following synusiae of plants (for further details see Chapter 4):

A. *Autotrophic plants (with chlorophyll)*

1. Mechanically independent plants
   - a Trees and 'shrubs'
   - b Forbs (dicotyledenous herbs)
   - c Graminoids (grasses and grass-like monocotyledons)
2. Mechanically dependent plants
   - a Climbers (vines and lianes)
   - b Stranglers
   - c Epiphytes

B. *Heterotrophic plants (without chlorophyll)*

1. Saprophytes
   - a Fungi and bacteria
   - b Vascular saprophytes
2. Parasites
   - a Non-vascular parasites
   - b Vascular parasites

Among these various synusiae can be recognised many types of commensalism (relationship in which one organism is benefited and the other unaffected), mutualism (in which both organisms benefit) and competition (which is harmful to one or both organisms).

At least one example of mutualism or symbiosis in the tropical forest deserves attention: the association of tree roots with fungi, commonly called mycorrhizae. Leaving aside the many cases where the fungus is found inside the root (endotrophic), which may often be a hidden form of competition, some comment is required about ectomycorrhizae, in which the rootlets of a tree are wrapped in a mantle of fungal hyphae. These have been shown by numerous studies in temperate forests to play an important role in the survival and the nutrition of trees, but they appear to be much less common in tropical forests. Among the families widely represented only some members of the Dipterocarpaceae and Caesalpiniaceae regularly possess ectomycorrhizae, plus a few tropical representatives of the Fagaceae, Pinaceae and Myrtaceae. In general, therefore, this relationship is fairly rare in the tropical forest, possibly because of the continued growth of roots or the relative scarcity of basidiomycetes in many parts of the humid tropics. It is interesting to note that some of the few ectomycorrhizal types occur in rather unfavourable soil or moisture conditions, for example in *Gilbertiodendron dewevrei* in Zaïre (Congo Kinshasa) (Peyronel and Fassi, 1957) and *Afzelia africana* in Ghana (Jeník and Mensah, 1967).

The diversity of animal life in the tropical forest is beyond the scope of this mainly botanical book, and only a few important points can be made. When walking in the forest one generally sees rather few animals, as numbers of them tend to live in the upper tree layers where food is most abundant. Amongst these species are many tree frogs, tree snakes and lizards, many mammals and countless arboreal insects and birds. Detailed observations have shown that amongst the animals of the tropical forest there is a vertical stratification, with particular species feeding, moving and resting at particular layers in the forest. Booth (1956) has shown this conclusively for monkeys in an African rain forest, while studies of bird behaviour patterns have been undertaken in numerous tropical countries.

The myriads of insects in the tropical forest may also occur in a stratified fashion, and they make many contributions to its ecology. Caterpillars, for example, may occasionally defoliate large portions of crowns, while termites form striking mounds. Ants are especially widespread, scavenging over the forest floor and the entire surface of many trees, and nesting in the soil and crown humus, on the trunks, in clusters of leaves or even inside the twigs and branches. They thus affect the nutrient cycle of the forest, especially those species which transport mineral soil and litter to build 'ant gardens' high up in the crowns.

A close association is found between ants and the tubers of some epiphytes, for example species of *Myrmecodia* and *Hydnophytum*, while certain trees such as *Canthium glabriflorum* in Africa almost invariably possess an ants' nest on their bole. Other trees, for example *Cecropia* spp. in South America, *Macaranga* spp. in South-East Asia and *Musanga cecropioides* in Africa often harbour ants in the pith of branches. The tremendous activity of ants in the tropical forest is probably stimulated by the presence of floral and extrafloral nectaries which are of widespread occurrence. In moving around they frequently act as pollinating agents (see section 5.8), and indeed, this phenomenon can be related to the evolution of cauliflory (see section 4.2, and Plate 8B). In addition, ants may act as seed distributors and also scavengers, removing rotting material and catching or driving off numbers of plant-sucking and leaf-eating insects (Gibbs and Leston, 1970).

As in other zones of the world, bees, wasps, flies, moths and butterflies also act as pollinators; however, birds and bats acting in this role appear to be limited to the tropics. Faegri and van der Pijl (1966) give many details of the adaptations found in 'bird-flowers' and 'bat-flowers', and similarly in 'flower-birds' and 'flower-bats'. Observing a flowering *Ceiba pentandra* just after sunset when numerous bats start visiting the opening flowers is one of the most striking spectacles of tropical ecology.

As mentioned above, there are relatively fewer ground animals in the tropical forest. Among the mammals the rodents are the most abundant and ecologically interesting, feeding on fruits and seeds which have fallen down to the forest floor, and contributing to the dissemination of plants. Several primates may also act in this way, while larger powerful mammals such as elephants may damage or destroy patches of the forest, thus affecting its regeneration and floristic composition.

Chapter 4

The forest community

As we have seen, tropical forests have in the past often been described as a confused tangle of vegetation of all kinds, making progress difficult or dangerous. The word 'jungle', originally meaning a low brush vegetation in Indian river valleys, became applied to any type of tropical woodland. It is certainly true that mixed tropical forests are very complex compared with the simple structure of single-dominant European beech or spruce forest. Yet the first impressions of travellers in the tropics may be misleading, since they are usually formed while passing along roads, railways, rivers and seashores, or walking along frequented paths and in the neighbourhood of villages. Consequently they see chiefly roadside thickets, riparian brush, seashore communities and disturbed secondary forest. However, such 'inaccessible jungle' can also develop under appropriate conditions outside the tropics; and conversely, it is important to realise that simpler structures tending towards the predominance of a single tree species occur in some tropical forests.

A reasonable comparison with temperate stands must involve investigating undisturbed mesic forests without abnormal drainage or nutrient conditions, in other words a comparison of the predominant climatic climaxes. There is adequate scientific evidence to show that in the humid tropics these forests are complex plant communities, though relatively well-organised in space, with distinct order in their vertical and horizontal arrangement. On the other hand, the edaphic climaxes, for example forests in the intertidal zone, swamp forests, peat forests and excessively-drained forests, can be almost monotonous in their uniformity. On the whole, the methods used in ecological studies of forests in temperate regions can be modified to suit tropical conditions, though there are some special difficulties.

The problem of comprehending tropical forests has been mentioned in section 2.1, and it is really only on freshly-cut margins that the structure and layering are clearly displayed. (See Plate 1.) It is surprising how quickly these become colonised by pioneer species, vines and coppice shoots (see Plate 6).

Special enumeration techniques are needed for serious ecological studies, to allow a complete *relevé* to be made of the spatial arrangement of individual trees, climbers, shrubs and herbs. Richards, Tansley and Watt (1939, 1940) developed a method of constructing 'profile diagrams' which has been successfully applied in many parts of the tropics. On a rectangular strip of minimum size 7·6 m by about 60 m, all the understory trees below an arbitrary size are cleared away. The position of the individual larger trees, their height, girth, shape and layering of the crown, are measured by appropriate instruments, and vertical profiles of the trees are drawn to scale on graph paper. If necessary, trees are felled in order to identify them on the spot or to collect herbarium specimens for further study. These profile diagrams (Fig. 4.1) can be supplemented by others showing the horizontal spread of stems and crowns.

Fig. 4.1 Vertical diagram of a sample plot in a tropical ombrophilous forest in the Ankasa Forest Reserve, Ghana. Width of the sample strip was 7·6 m. (After Jeník and Hall.)

Experience has shown that the semi-quantitative estimates commonly adopted in the analysis of plant communities in temperate countries, for example the methods of the Zürich-Montpellier School, encounter difficulties when applied to mixed tropical forest. For instance, many tropical trees cannot be identified from a distance, and so complete enumerations or phytosociological records are essential. There have been some recent advances in the application of quantitative methods in the Solomon Is. (Greig-Smith *et al.*, 1967), Sabah (Austin and Greig-Smith, 1968) and Ghana (Lawson *et al.*, 1970).

The current taxonomic knowledge of plants in tropical regions is another problem in ecological studies. Although detailed floras exist or are in preparation for most countries, reliable keys for identification are seldom available. It often happens that scientific keys are less useful than the knowledge of local 'tree-spotters' and foresters who can distinguish many hundreds of species with great skill.

The ecologist studying the tropical forest is concerned both with the spatial arrangement of trees and with the inter-relationships between the plants and the physical environment. In ecology causal relationships are anticipated between the soil and climate on the one hand and the plant community on the other, and one assumes that a certain composition of the forest is controlled by particular factors of the environment. However, in mixed tropical forests, as we have seen, the soil and microclimate are also very greatly modified by the vegetation. Moreover, large numbers of species appear to be 'interchangeable', occurring apparently at random over an area in their particular ecological niche, presumably because several species have very simiar physiological and ecological requirements. In such circumstances a statistical approach is more relevant than a purely functional one. Furthermore, an evolutionary standpoint is desirable in assessing the different compositions of tropical forests (see section 4.4).

4.1 Vertical and horizontal structure

The tropical forest is a three-dimensional phenomenon, and therefore the position of any elements in its organisation, plant organs, individual plants, populations, plant synusiae, etc., can be properly described only by a system of co-ordinates. Since the trunks are generally vertical, it is convenient to take as a basis this plane and a horizontal one at right angles to it. Accordingly one may speak of the vertical or the horizontal structure of the forest.

Amongst the features of the vertical structure, four characteristic points deserve attention:

1. The height achieved by the forest
2. The above-ground layering
3. The root-mass layering, and
4. The maximum depth reached by root systems.

Although very large, the biggest emergents of the tropical forest are not quite as tall as, for example, the redwoods of California or some of the gum trees of subtropical Australia. The maximum height of the canopy in tropical forests seldom exceeds 50 m, though exceptional specimens may reach 70 or 80 m. As shown in Figs. 2.3, 2.4 and 2.5, the general height of tropical forests is controlled partly by the total annual rainfall, seasonal periodicity of rain, soil drainage and amount of nutrients. On the whole, the tallest structures are found in areas with an alternation of humid and drier seasons, rainfall totals of about 2 000 mm and freely-drained soils. These conditions presumably favour in various ways the growth and longevity of the trees. By contrast, high-rainfall areas without marked seasonal changes and sites with impeded drainage often tend to give smaller trees, sometimes with a shorter life-span. Other adverse environments which may reduce the forest height are found, for example, on wind-swept slopes and hilltops, seashores influenced by salt-laden winds and in mountainous areas. The most luxuriant trees and stands of all, sometimes exceeding 60 m, are to be found in the upper free-drained parts of flood-plains of rivers, where flooding has introduced fertile silt and no form of water stress occurs.

In the Amazon Basin Hueck (1966) found the average height of the forest to be 30–40 m, with *Dinizia excelsa* (Leguminosae) and the Brazil Nut, *Bertholletia excelsa* (Lecythidaceae) as examples of species where individuals may reach 50 m. In tropical Africa similar size ranges are encountered, and a gradient of increasing forest height can be recorded along an ecotone from high-rainfall areas (stands approximately 30 m high) towards evergreen seasonal forest (approx. 40 m). *Entandrophragma cylindricum* (Meliaceae) and *Piptadeniastrum africanum* (Leguminosae) are examples of giant trees which occasionally protrude far above the average level of the canopy. In the Indo-Malaysian Tropical-Forest Region, *Dryobalanops aromatica* and other species of Dipterocarpaceae can reach 60 m, and the tallest tree ever recorded in the tropical forest grew here: a specimen of *Koompassia excelsa* (Leguminosae) which measured 84 m (Foxworthy, 1927).

The effect of altitude on the vertical structure of tropical forests can be clearly seen from the data in Table 4.1.

Table 4.1. Relationship between altitude and the height of the forest canopy in the Malayan Peninsula. After Robbins and Wyatt-Smith (1964).

Forest formation	*Altitude (m)*	*Height (m)*
Lowland dipterocarp forest	150	42
Upper dipterocarp forest	800	30
Lower montane oak-laurel forest	1 500	21
Montane ericaceous forest	1 800	15

The top of the canopy in tropical forests is much more undulating than is usual in temperate broad-leaved forests. Profile diagrams and photographs (see Fig. 4.1 and Plate 1) show a discontinuous upper-tree layer of emergents which have a particularly distinctive appearance when viewed from an aeroplane. Large areas of untouched forest have for this reason been somewhat inelegantly referred to as 'cabbage patches'.

Most species, of course, never become emergents, and do not exceed 20–25 m even at maturity. The taller ones belong to the middle tree layer, which is more continuous, while lower down in the under-story there are two classes of smaller trees: some seedlings and saplings which may later reach the higher levels, and other 'pigmy trees' which stay permanently in the undergrowth.

Interpretation of the term 'layering' has caused considerable controversy amongst ecologists. Some hold that marked vertical stratification is very characteristic of the tropical forest, while others state that no clear-cut layering can be found. The reasons for this disagreement lie in how strictly the term 'layer' is defined. In its full sense it includes the idea of clearly-defined levels reached by different sets of crowns and roots. In this respect, there are no layers in primary tropical forest, except occasionally when many seedlings of the same age grow up together in a gap. However one may commonly use the term 'layering' to imply that tree crowns and roots achieve widely

Table 4.2. Theoretical layering in tropical forest

Layer	*Brief description*
Upper tree layer	Emergent trees, woody climbers and epiphytes above 25 m
Middle tree layer	Large trees and woody climbers from about 10 to 25 m
Lower tree layer	Small trees and saplings reaching about 5 to 10 m
Shrub layer	Tree seedlings, shrubs, *krummholz*, small pigmy trees from 1 to 5 m
Herb layer	Smaller tree seedlings, forbs, graminoid plants, ferns and bryophytes up to 1 m
Upper root layer	Compact root-mass in the surface soil down to 5 cm
Middle root layer	Less abundant tree roots in the subsoil from about 5 to 50 cm
Lower root layer	Scattered tree roots below 50 cm

Data based on West African ombrophilous and evergreen seasonal forest.

different heights and depths in the total vertical profile. In this sense, of course, the complex mixed structure of the tropical forest is a prime example of layering, and the term 'layer' has also proved itself to be very useful in ecological descriptions. (See Table 4.2.)

The trees, chief components of the vertical structure, will be discussed in section 4.2. The other members form a much smaller part of the biomass, but they contribute in a unique fashion to the layering of the tropical forest.

Amongst them are the epiphytes, which in accordance with Tixier (1966) can be divided into macro-epiphytes (vascular plants), and micro-epiphytes (mosses, liverworts, lichens and algae). Rather a limited number of families of angiosperms is represented, particularly common being members of the Orchidaceae, Bromeliaceae, Cactaceae, Asclepiadaceae and Rubiaceae (Plates 27 and 28); and many genera of ferns also form macro-epiphytes. Bryophytes can be very abundant, and in a study of epiphytes on the southern slopes of the South Annamitique Range in Vietnam, Tixier (1966) found 116 species of mosses, 110 of liverworts, and also 36 species of lichens.

According to their micro-habitat, epixylous epiphytes, covering branches, limbs and trunks, can be distinguished from epiphyllous types which grow on the surface of living leaves (Plate 11A). Liverworts,

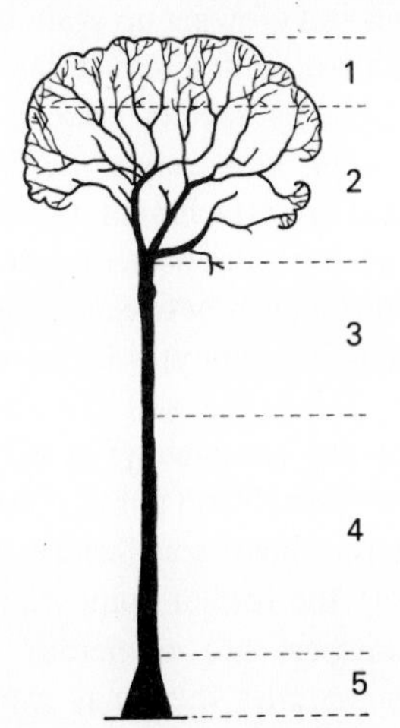

Fig. 4.2 Schematic division of microhabitats of epiphytes within an emergent tree in the ombrophilous forest: 1 – fully-exposed terminal twigs of the crown with micro-epiphytes; 2 – main zone of epiphytes covering large limbs; 3 – drier upper part of trunk with crustaceous lichens; 4 – moister lower part of the bole with lichens and frequent bryophytes; 5 – base of the tree, with numerous shaded pockets between buttresses and spurs, covered by bryophytes.

lichens and algae all occur in the epiphyllous synusia, and this peculiar phenomenon has not so far been recorded outside the tropics.

Epiphytes grow, of course, under very diverse conditions according to their exact position in the forest structure, and a classification of these micro-habitats is given in Fig. 4.2. Sudden transitions can occur, for example if the zone 2 epiphytes, which normally experience semi-arid conditions with full radiation (Plate 5), fall down into the moist, shady undergrowth on a broken limb or uprooted tree. Such an abrupt ecological change has few parallels in other plant communities, and it may have been a factor in the evolution of rich tropical floras (section 4.4).

Climbers of various sizes, morphology and life-form add to the complexity of the vertical structure. The woody climbers or lianes can become very large, with thick convoluted stems (Plate 8A). These can show as many as seven different types of abnormal cambial activity (Obaton, 1960), which result in an irregular outline of the stem and even occasionally in a rope-like appearance due to the partial disintegration of the tissues. At different times in the life of a liane, its growing points and foliage can be anywhere from the herb layer up to the top of emergents, as it climbs upwards on trees of various sizes. It may later slip or fall down if its support should rot or break, forming great loops on the ground and growing up again by another route.

The smaller climbers are more tolerant to shade in the forest interior and include many herbaceous species (vines), such as some of the Araceae. Both woody and herbaceous climbers can be classified according to their organs of attachment to the supporting tree: for example stranglers, twining climbers, root-climbers, and tendril-climbers. One can also distinguish between climbers germinating in the soil and those which start as epiphytes in the crowns and send down roots which eventually reach the soil. In the latter group is the remarkable life-form of the strangler (Fig. 4.3 and Plate 17A). The descending roots thicken, branch and anastomose, ultimately forming a compact casing around the 'host' tree, which is then overgrown and literally strangled, leaving the former epiphyte as an independent tree on its own roots. Stranglers are numerous in the genera *Ficus*, *Schefflera* and *Clusia*; sometimes they may even become members of the upper tree layer.

In closed ombrophilous lowland forest, the herb layer seldom covers more than 10 per cent of the surface, and is comparable in extent with the typical undergrowth of European beech forests. Moreover, there are generally fewer species of ground herbs than woody plants. For example, in an area of a few square kilometres of primary forest in

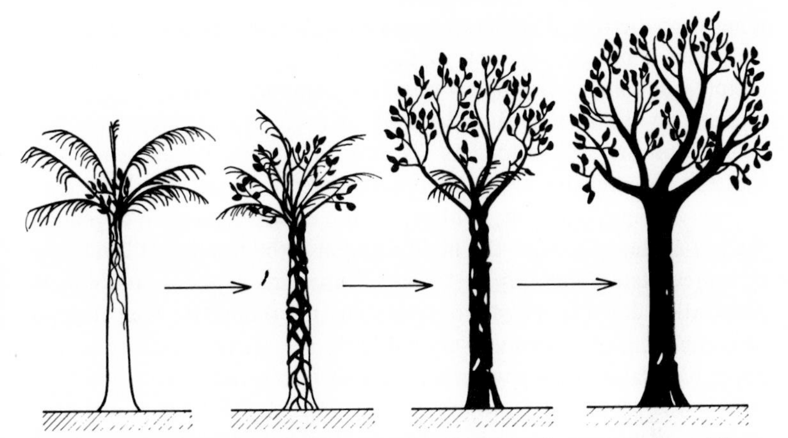

Fig. 4.3 Four stages in the establishment of a strangler. *Ficus leprieuri* on *Elaeis guineensis*.

Guyana the herb layer comprised less than 30 species of flowering plants, 10–20 species of ferns, plus a single species of *Selaginella*, as compared with several hundreds of species of large woody plants (Richards, 1952).

Dicotyledonous families commonly represented in the herb layer include the Rubiaceae, Gesneriaceae, Begoniaceae, Melastomataceae and Acanthaceae, while monocotyledons often belong to the Cyperaceae, Gramineae, Commelinaceae, Marantaceae, Zingiberaceae and Araceae (see Plate 9AB). A number of the ground herbs show interesting features of shape, colouring and surface structure which have made some of them popular house plants, for example many species of *Begonia*. Grasses and sedges growing on the forest floor often have very broad leaves unlike 'typical' representatives of these families, and some dicotyledonous herbs have a series of regular bumps on each side of the main vein which may play a role in surface drainage comparable to drip-tips (see section 3.3 and Plate 9A).

Genuine shrubs are seldom found in the untouched rain forest. The majority of the small woody plants scattered in the undergrowth are either seedlings of bigger trees, palms or 'pigmy' trees. The last mentioned have a distinct, often unbranched axis, with a small head of large leaves, and they remain as trees *en miniature*, seldom reaching a height of more than 5 m (Plate 3). In the West African forests *Anthonotha macrophylla* and *Scaphopetalum amoenum* develop large vegetatively-spread thickets which can be classified as a special life-form known as *krummholz* (Jeník, 1969). In both species young woody stems curve downwards and root in the soil, after which lateral buds lying near the top of the large loops grow out and behave in a similar way. This is but one example of the fascinating physiological problem of apical dominance in tropical trees (see Damptey, 1964).

Data on the underground layering of roots and the depth of root penetration are still scarce. As discussed in section 3.4 and shown in Fig. 3.9, the available evidence suggests that the upper root layer may be very distinct, and that, contrary to the casual impression of general shallow-rootedness, both the middle and also the lower root layer may be represented. Some tropical trees can form large tap-roots and sinkers penetrating down to a depth of several metres, though the root systems of climbers, pigmy trees and ground herbs are usually confined to the upper root layer. In the water-logged soils of alluvial, swamp and peat forests, the prevailing depth of rooting is generally shallow, deeper root systems apparently occurring only in connection with special root adaptations, such as peg-roots and knee-roots (see section 4.2).

An analysis of the horizontal structure of tropical forests is a very complicated procedure, and has to be done on a large enough area because the effective extent of many populations is so great. There is also much variety in groupings of trees, density of undergrowth, location of seedlings and in patterns of populations. Another important consideration is the occurrence and horizontal distribution of natural gaps and other natural forest margins. Many of these questions are discussed further in sections 4.3 and 4.4, but it may here be noted that there is some pattern in the apparent confusion and disorderliness of the horizontal arrangement.

Many observations and measurements have shown that this horizontal structure commonly forms a mosaic of three phases corresponding to the three light regimes described in section 3.1. These do not remain static but change with time. The mosaic consists of:

1. The phase of the mature stand (corresponding to the dim phase of light). This is the most extensive and stable part of the mosaic. Under the unbroken canopy of the larger trees, seedlings and pigmy trees are scattered irregularly, with lianes and especially vines rather rare. Grasses and forbs do not cover more than a small proportion of the soil. It is usually easy to walk around (see Plates 3 and 4).

2. The phase of the open gap (corresponding to the light phase). In virgin forests gaps are caused by the falling down of dead trees and the loss of parts of larger individuals. Natural decay by fungi, destruction by insects or mammals, and windthrow by hurricanes are the usual factors causing such openings. The interruption of the canopy stimulates luxuriant development of climbers and ground herbs, and encourages faster growth of tree seedlings. It is likely that it also leads to the germination of dormant seeds (see section 5.9). Unless there is abnormal wind action, the area of these natural gaps usually represents less than one-twentieth of the entire forest. In swamp forests this phase is more extensive because the trees tend to die earlier: large forbs, grasses, ferns and palms spring up most frequently in these gaps.

3. The phase of the dense thicket (corresponding to the dark phase). On the forest floor beneath the canopy, thick, tangled masses of living and dead climbers and branches lie where they fall, and in addition, the phenomenon of *krummholz* creates similar conditions. The already low light intensity is further reduced and in these dark conditions there is virtually no herb layer and no natural regeneration of trees. This is generally less common than the phase of the open gap, unless *krummholz* is locally abundant.

Forest margins represent another interesting feature in the horizontal structure of the tropical forest. Natural margins are formed along riversides, on the seashore and around large clearings caused by hurricanes. Artificial edges to the forest occur, for example, along roadsides, around compounds and villages, and on sharp fire-induced boundaries between forest and savanna. These marginal communities are rather unlike those in the interior of the forest. With more small trees, vines, woody climbers and heliophilous ground herbs a dense strip or patch is created (see Plate 6). The luxuriance of this marginal thicket is presumably due both to increased light intensity and decreased root competition, and may sometimes be enhanced by nutrient release, river fogs and spray. Birds, small mammals and insect colonies may also contribute by their activity to the different constitution of the forest margins. How quickly and efficiently such a thicket appears can be easily observed along newly constructed roads. In the Atewa Range in Ghana, for example, abundant regrowth of *Musanga cecropiodes* and *Trema guineensis*, sometimes with the razor-sharp climbing sedge *Scleria boivinii*, appear in the first year. Within five years, thick stands of trees more than 10 m in height and 30 cm in girth have closed up both edges of the primary forest (see also Fig. 27 for growth rates of pioneer trees). One may wonder in fact where all these new trees come from, since mature specimens bearing ripe fruits of such pioneer, light-demanding species are rather uncommon in the closed forest. Observations in Nigeria have shown, however, that seeds of some species may be accumulated in the soil over quite long periods, remaining viable but dormant (Keay, 1960; see also section 5.9).

4.2 Features of tropical trees

Trees being the dominant organisms in tropical forests, their life-form affects the general physiognomy, primary production and overall life-cycle of the community. Many characteristics of tropical trees do not differ from those in other parts of the world, whereas there are certain features of branching habit, foliage, flowers, fruits and root systems which are seldom or never encountered in other geographical zones. It is to these that the following pages chiefly refer, but first a general account of root behaviour will be given, since this is less well known than that of the shoots.

In describing the form and growth of roots, one has to bear in mind both ontogenetic changes and the interaction between heredity and environmental factors. Thus one can distinguish between the form of roots which are restricted by a hard lateritic pan or by a permanently impeded horizon from the root system which might have developed had they been growing in deep and fertile soil. Moreover, the growth patterns of large skeleton roots are likely to have become very different from those of distal end-roots.

In older dicotyledonous trees, four types of root systems can be tentatively classified according to the position and shape of their skeleton roots (see also Coster, 1932, 1933; Wilkinson, 1939).

1. Root system with thick horizontal surface roots, frequently merging into large spurs or buttresses; with weak vertical sinkers and tap-root entirely absent (Fig. 4.4).
2. Root system with thick horizontal surface roots, and well developed tap-root and sinkers.
3. Root system with weak surface roots, and a rich system of many oblique 'heart' roots and a prominent tap-root.
4. Root system with numerous sizable aerial stilt-roots, and a network of weaker underground roots (Fig. 4.4).

Among the emergent trees, types 1 and 2 clearly prevail, and type 3 is most frequent in the smaller trees and big woody climbers. Type 4 is most characteristic of trees growing in waterlogged sites, although species like the African *Uapaca guineensis* and *Xylopia staudtii* (Plate 10) develop stilt-roots even on mesic soils. The size, shape and position of skeleton roots also vary according to the age of the tree. Some type of tap-root is produced in nearly all seedlings but other roots may later become predominant (see Mensah and Jeník, 1968).

Aerial stilt-roots (Jeník, 1973) are prominent examples of a special feature of tropical forest trees, and have therefore frequently been

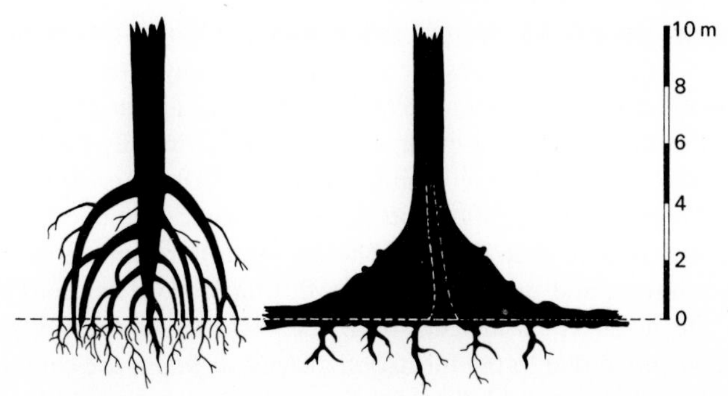

Fig. 4.4 Stilt-roots of *Uapaca* sp. (left), showing a root system of type 4; and heavy buttresses of *Piptadeniastrum africanum*, illustrating a type 1 root system.

photographed. However, many other tropical trees develop some part of their root system above the ground. There is a great variety of these roots including small spur roots, large buttresses, and numerous kinds of pneumorhizae. All these features provide valuable characters for identification in the field (see for example Schnell, 1950).

Buttresses can vary in size, shape and number; at their least prominent being merely a small spur or swelling at the base of the stem, not unlike those found in temperate conifers. At the opposite extreme they may develop as massive structures reaching as much as 10 m up the stem and much further in the horizontal plane. In such cases they provide a considerable obstacle to felling, and on occasion a platform has to be built. Examples of emergent trees possessing very large buttresses are *Mora excelsa* in South America, *Kostermansia malayana* in South-East Asia and *Piptadeniastrum africanum* in Africa.

On some trees the buttresses are very thin and flat and may sometimes be suitable as boards for house building. Others are very twisted, and anastomose providing unusual microhabitats. From the morphological point of view they are, of course, part root and part stem. The function of buttresses of different types is not clearly understood, although many contradicting theories have been offered. It seems likely, however, that some of them represent adaptations increasing the stability of big trees.

In the lower portion of the trunk of some species it is normal for adventitious roots to be formed. Sometimes these may remain small and thin, never reaching the ground (Plate 13A), while in other cases the elongation and secondary thickening are stronger, and when this kind have penetrated the soil they form stilt-roots. Associated with these stilt-roots, for example in a few palms and in *Bridelia* spp.,

Macaranga barteri, *Klainedoxa gabonensis* and *Commiphora fulvotomentosa*, unusual root-spines are formed which may later develop into full stilts (Jeník and Harris, 1969). In some cases a great proportion of the root system is above ground, and the tree is virtually held up only by the stilt-roots (see type 4 above and Plate 10). It is clear that in many of these cases the stilt-roots have a mechanical function, and it is interesting to note that they tend to be more frequent on the softer soils of swamps and mangrove woodland. Considerable tensions are set up in the aerial roots hanging from the branches of the subtropical *Ficus benjamina* due to the formation of tension wood, the contraction of which could add to stability of the tree (Zimmerman *et al.*, 1968).

In these same impeded soils, various less conspicuous and lesser-known features of tropical roots occur, notably various types of pneumorhizae, which grow upwards from below the soil level. For instance, *knee-roots* (Jeník, 1967) consist of small loops of portions of roots which leave the soil and then enter it again. Some are formed from laterals which grow up from a deeply situated horizontal mother-root as found in *Symphonia globulifera* in South America and in *S. gabonensis* and *Mitragyna ciliata* in Africa (Fig. 4.5(A)). But there is another 'serial type' where a horizontal root curves in and out of the soil, for example in the mangroves *Bruguiera* spp. and *Ceriops* spp. (Fig. 4.5(B)). These knee-roots should not be confused with the 'root-knees' of the *Taxodium* type which result from local intense cambial activity on the top side of a subterranean root (Fig. 4.5(C)).

Peg-roots are a very remarkable type of root modification which also arise from a horizontal mother-root. Unlike knee-roots they grow more or less vertically and can even reach a height of 30 cm. They are often

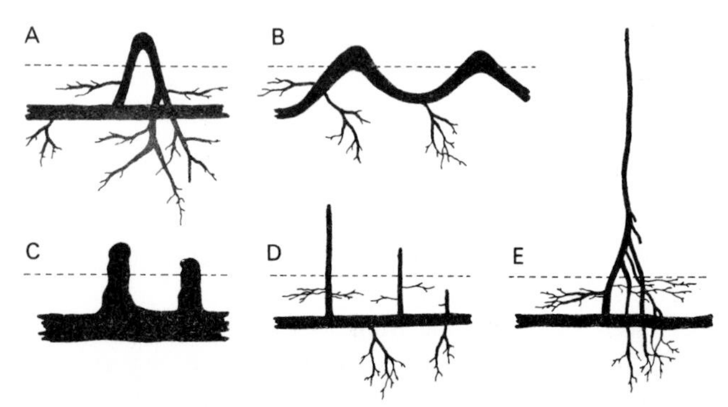

Fig. 4.5 Various types of pneumorhizae occurring in impeded tropical forest soils: A – lateral type of knee-root; B – serial type of knee-root; C – root-knees of the *Taxodium*-type for comparison; D – peg-roots; E – stilted peg-root.

about the thickness of a pencil, and look rather like pegs inserted into the soil (Fig. 4.5(D) and Plate 13B); they are well-known in *Avicennia*-mangrove swamps all over the tropical world. Peg-roots are also formed in freshwater swamp (Ogura, 1940), for example in palms such as *Raphia* spp. and *Ancistrophyllum* spp. and in certain dicotyledonous trees, notably *Anthocleista nobilis* and *Voacanga thouarsii* (see Jeník, 1970a and 1971b). In species of the mangrove *Sonneratia* the peg-roots show strong secondary thickening and form cone-shaped organs protruding above ground. In this and the other dicotyledons, the root surface is covered by numerous lenticels.

Pneumathodes are rather similar, but here only the whitish root-apex protrudes above the soil surface. Loose powdery cortex is formed just below the tip as can be observed in several species of *Phoenix* and in *Laguncularia racemosa* (Jeník, 1970b). The latter species does not form true peg-roots as is often stated.

Another special type is the *stilted peg-root* which appears to be a secondary adaptation of this kind of pneumorhiza. In *Xylopia staudtii* in tropical Africa, some of the slender peg-roots may continue growth, twisting upwards and reaching as high as 2 m. These later branch, forming a set of lateral roots which grow down again and anchor themselves into the soil (Fig. 4.5(E)). This set of stilts occurs in addition to the ordinary adventitiously formed stilt-roots (see Jeník, 1970c).

Many of the preceding types of aerial roots have been called ‘breathing roots’ or ‘pneumatophores’, but in view of other uses of this word it is preferable to regard them as different kinds of pneumorhizae. It is quite likely that many or perhaps all of them are adaptations to growth in soils lacking aeration, and Chapman (1944) has shown that gaseous exchange takes place very freely in *Avicennia* through the lenticels and large intercellular spaces. Further evidence is provided by the absence of some root modifications when *Xylopia staudtii, Mitragyna ciliata* and *Anthocleista nobilis* are growing on mesic sites. In the latter species, examples have been found where peg-roots were formed on a mesic site but only at the bottom of a man-made ditch. Another interesting point is that some pneumorhizae, in common with many aerial roots of epiphytic orchids, contain chlorophyll and may therefore make some contribution to their carbohydrate content.

The shapes and branching habits of tree trunks and crowns are far more variable in the tropics than in temperate regions (Corner, 1940). A recent classification of form and morphogenesis in young tropical trees has identified as many as 21 types or ‘models’ even though only about 25 per cent of the African species were taken into account (Hallé and Oldeman, 1970). Many palms, tree-ferns and a few dicotyledons have a single stem which is normally unbranched throughout life (Fig. 4.6(A)). These are all 'palm-like' trees, but in addition the juvenile stages of a number of others, for example some Meliaceae and Dipterocarpaceae, are unbranched for a number of years (see section 5.7).

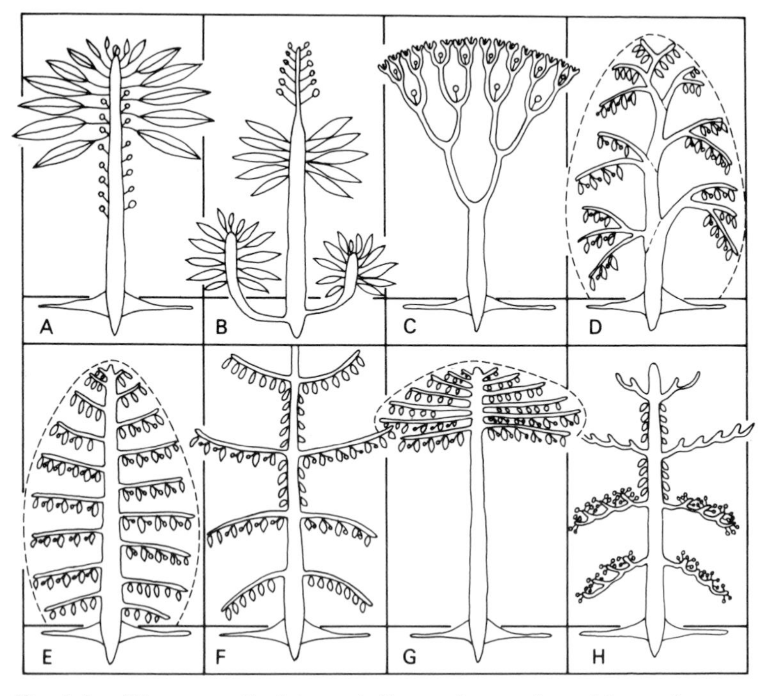

Fig. 4.6 Diagrammatic interpretations of some branching and reproductive patterns in tropical trees, according to the classification of 'models' by Hallé and Oldeman (1970):

Unbranched trees:

A – Model of Corner; e.g. *Elaeis guineensis, Mauritia flexuosa, Cocos nucifera, Pycnocoma angustifolia, Phyllobotryum soyauxianum, Ficus theophrastoïdes.*

Branched trees:

branches all of equal status:

B – Model of Tomlinson; e.g. *Raphia gigantea, Euterpe oleracea.*

C – Model of Leeuwenberg; e.g. *Manihot esculenta, Plumeria acutifolia, Rauvolfia vomitoria, Anthocleista nobilis, Alstonia sericea.*

sympodially-formed trunks and branches:

D – Model of Troll; e.g. *Delonix regia, Piptadeniastrum africa-*

The great majority of trees branch sooner or later and can be divided into three types: firstly those where the branches are all of equal status (see Figs. 4.6(B),(C)), secondly where there are clearly different orders of branches (see Figs. 4.6(E)–(H)), and the third type where the main stem consists of a succession of sympodially growing axes whose terminal parts curve over and form branches (Fig. 4.6(D)).

The crown shape of a mature tree depends partly on which of these categories it belongs to, but also on its adult characteristics and the environments in which it has grown. Some have enormous near-horizontal limbs up to 20 m in length (Plate 5), occasionally even showing buttresses where they join the trunk. In Plate 2 various less massive branching types can be seen, and these together with many other forms make the forest structure extremely variable.

The majority of tropical trees possess a smooth, light-coloured bole of a cylindrical shape. Some, however, show features such as fluting, spiral twisting, large spines as in *Fagara macrophylla* or *Erythrina mildbraedii* (utilised as 'rubber stamps'), adventitious roots such as those in Plate 13A, etc. These can be very useful in identification, as can the shades of bark colour, though this can be influenced by the crustaceous lichens present. Ridged and furrowed bark or rhytidome is not nearly as common as in temperate trees, but features of the bark are very distinctive when displayed in a 'slash', cut through the outer and inner bark to the young sapwood. Characteristics such as its colour, texture, odour and taste, and also the presence of latex and various other exudates, can be used by experienced foresters, tree spotters and botanists for rapid recognition in the field. The anatomical differences which underlie the slash characters have for the most part not been

---

num, Cassia javanica, Parinari excelsa, Pterocarpus officinalis. (Also the European tree, *Tilia platyphyllos*.)

branches of different status:

E – Model of Rauh; e.g. *Hevea brasiliensis, Pentadesma butyracea, Entandrophragma utile, Triplochiton scleroxylon, Cecropia peltata, Artocarpus incisa, Musanga cecropioides, Khaya ivorensis.* (Also many common European trees; e.g. *Quercus, Fraxinus*.)

F – Model of Massart; e.g. *Ceiba pentandra* (see Plate 18), *Diospyros matherana, Pycnanthus angolensis, Anisophyllea* spp.

G – Model of Cook; e.g. *Phyllanthus mimosoides, Panda oleosa, Canthium glabriflorum, Glochidion laevigatum.*

H – Model of Aubréville, otherwise known as 'pagoda-trees'; e.g. *Terminalia catappa, T. ivorensis,* (see Plate 23A), *Sterculia tragacantha, Omphalocarpum elatum, Manilkara bidentata.*

studied scientifically (but see Whitmore, 1962a, b); in some countries the local knowledge of barks for medicinal purposes and poisons may be very advanced. (Indiscriminate tasting of bark is not advisable, therefore.)

New branches are only exceptionally produced from dormant epicormic buds on older trunks, though this is quite common in one or two swamp species such as *Protomegabaria stapfiana* and *Grewia coriacea* in tropical Africa. Vegetative reproduction occurs in a few dicotyledons, such as *Scaphopetalum amoenum*, and in a number of monocotyledons, for example *Pandanus* spp. and bamboos. Under experimental conditions at least, *Chlorophora excelsa* will regenerate buds from detached roots.

The terminal and lateral buds on tropical tree shoots are often smaller than in temperate regions, and frequently lack clearly defined bud-scales, though these tend to be more common in flower buds. In some cases the dormant growing-points may be surrounded by stipules, hairs or resinous secretions, or by a small pocket in the leaf-joint or pulvinus (see Plates 7B and 21B).

Some of the attractive features of young leaves are discussed in section 5.4. Mature leaves vary very considerably although the visitor to the tropical forest is often struck by the dominance of dark green, oblong-lanceolate, entire and leathery leaves (or leaflets) which make the foliage within the forest appear very uniform. A closer study shows that even in the undergrowth layer there is a variety of shapes of leaf-blades, sufficient to be sometimes of use in the identification of trees. A striking feature of the majority is the presence of drip-tips, the significance of which was mentioned in section 3.3. The quicker drying will have the effect of enhancing transpiration, and it is possible that it also reduces the tendency for epiphyllous plants to colonise and mask the leaf surface (see Fig. 4.7 and also Plates 7A and 11A).

The emergent trees with their crowns exposed to strong insolation and intermittently drying conditions generally have much smaller leaves without a distinct acumen. Some of these are capable of various movements with regard to the sun's position and the time of day (see section 5.4). The position of the leaflets in many leguminous trees, for example *Piptadeniastrum africanum*, alters so much that around midday the light intensity in the lower stories is substantially increased. These movements are brought about in many tropical leaves by specialised pulvini or leaf-joints (see Plate 7B). Another feature is the tendency of riverside species to develop linear-lanceolate leaves.

The patterns of leaf-shedding and flushing of new leaves will be discussed later in Chapter 5. These changes in the canopy are obviously

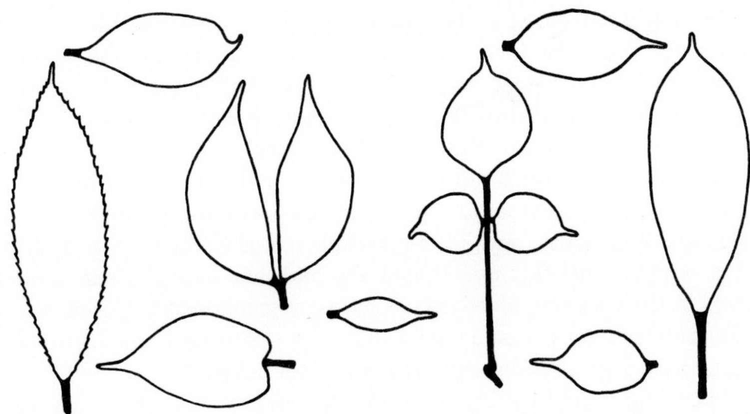

Fig. 4.7 Various shapes of leaves with conspicuously-differentiated drip-tips.

of basic importance to the life of the forest and determine the general physiognomy and classification of tropical forests (see section 4.5).

The diversity of flowers, fruits and seeds is remarkable, with a great variety of generative organs and many pollination and dispersal mechanisms. Detailed description of flower and fruit morphology, so fundamental in the classification of plants, is outside the scope of this book, but may be obtained from local floras. Some experimental details on flowering and fruiting will be described in sections 5.8 and 5.9, while a few unusual characteristics are mentioned here.

An interesting feature is the occurrence of flowers and inflorescences (and consequently fruits) on leafless woody trunks and larger branches, which is known as 'cauliflory' (Plate 8B). This phenomenon in its broad sense is subdivided in Table 4.3 and discussed further by Richards (1952).

Table 4.3. Cauliflory (broad sense) and its subdivision

Cauliflory	*Flowers on the trunk, leafless twigs and roots*
Ramiflory	Flowers on larger branches and leafless twigs, but absent from the trunk
Trunciflory	Flowers on the trunk, but not on the branches and twigs (equals cauliflory in the narrow sense)
Basiflory	Flowers at the base of the trunk
Flagelliflory	Flowers on pendulous twigs spreading down from the lower part of the trunk on to the ground surface
Rhizoflory	Flowers on the roots

A very striking case is the positioning of the flowers on the leaves, which we suggest may be termed 'phylloflory'. This is shown in several genera of Flacourtiaceae, as for example in the pigmy tree *Phyllobotryum soyauxianum* (Hutchinson *et al.*, 1954), where the inflorescences are borne on the midrib of the assimilating leaves.

Cauliflory and phylloflory, like other adaptations in floral biology, can best be understood in terms of the predominant pollination agencies. Within the forest the air movement is generally very slow, and it is usually found that insects, and also birds and bats play important roles in the transport of pollen. As has been mentioned in section 3.6, ants are constantly passing along the trunks, branches and leaves of many trees and can easily serve as a vehicle for pollen.

With regard to fruit and seed dispersal also, a large proportion of trees in tropical forest appear to be adapted to animal activity. Even in the upper tree layer only about half of the species are dispersed by wind, the majority of the rest being distributed by birds, rodents, monkeys and insects. In the middle and lower tree stories animal agents prevail (Jones, 1956). Quite often the fruits are heavy, for example *Omphalocarpum* spp., and perhaps are therefore more likely to reach the soil surface (Plate 14B). Fleshy fruits attract animals which spread the seeds throughout the forest. A number of trees have explosive fruits dispersing the seeds, and an unusual example occurs in *Allexis cauliflora* (Plate 8B), an undergrowth tree which lobs its single seed over considerable distances (Jeník and Enti, 1968). Seeds and fruits of wind-dispersed trees often have wings or hairs and some can glide in almost still air.

4.3 Loss and replenishment

While describing the structure of tropical forests and considering the morphology of its component trees, it has sometimes been necessary to neglect the constant changes which take place. Actually, of course, tropical forests are dynamic communities with a continuous sequence of events and ceaseless development. Seeds germinate, seedlings spring up and many die, saplings and poles enlarge, large trees spread their branches widely and enter the reproductive phase, old trees cease growing and die off, dead trunks break down and the fallen wood decays. At the same time individual species and populations compete and combine, interacting in the life of the whole ecosystem.

In temperate regions successive changes in the forest are clearly recorded in the annual growth-rings in the stem. Both the age and the growth rates of the trees, as well as their past and present interdependence can be traced with a high degree of accuracy from the cut stump. In the tropics the forester and plant ecologist are generally without this reliable internal 'calendar', since the majority of trees do not form clear annual rings (see section 5.6).

Available information on the probable ages reached by tropical trees is still scarce. Data from the Indo-Malaysian dipterocarp forest suggest an average maximum age of the emergent trees of 200–250 years. Jones (1956) used several methods to estimate tree age in Nigeria, and believes that the largest trees of *Lophira alata* and *Guarea cedrata* can be 300–350 years old. The size of the tree cannot serve as a reliable indicator of its age unless successive measurements on sample plots allow reasonably accurate extrapolation. Individual emergents of similar girth may vary markedly in age, for example *Triplochiton scleroxylon* was estimated to reach the chosen diameter of 102 cm in 50 years, while *Khaya ivorensis* took about 115 years, and *Guarea thompsoni* 345 years.

As with most woody plants, the death of old trees in tropical forests generally does not occur suddenly, as an abrupt stopping of meristematic activity throughout the whole plant. The final stages of tree senescence include the death of parts of the root system, reduced cambial activity and shoot elongation, stagnation and gradual death of branches, and increasing decay of heartwood in the centre of the trunk and bigger limbs. The weakened organism is attacked by insects, fungi and bacteria, and at this stage one may speak of the 'death' of a tree. Later the crown breaks down limb by limb, and eventually the whole trunk collapses. More sudden can be the end of a tree destroyed by a hurricane, uprooted by an elephant, or torn down by another tree falling nearby. While camping in the tropical forest one may hear occasional sounds of breaking timber at any time of day or night, and these are a useful reminder of the continuing changes in the forest structure.

The natural gap left by the dead tree represents a vacancy in the canopy and rhizosphere that is available to new competitors. The size of the gap may often be enlarged because of damage to the neighbouring trees from falling limbs and trunks, together with attached climbers. The former equilibrium is greatly disturbed, with increased insolation reaching the forest floor, changed spectral composition of light, marked fluctuation of soil and air temperatures, more rapid release of nutrients and reduced root competition. Usually these environmental changes are quickly followed by the germination of numerous seeds in the soil, and by promotion of the growth of stunted seedlings and saplings. In addition, pioneer herbaceous and woody species appear rapidly and within a few years the gap has been filled with vigorous regrowth.

In this way generation proceeds over great areas which are not influenced by abnormal destructive factors. In regions affected by hurricanes, landslides, volcanic eruptions, fires or the activity of elephants, the openings in the forest may be much larger, even covering square kilometres. This creates extensive natural secondary forests, while recent exploitation by man has made artificial secondary forests a frequent or prevailing formation in many accessible regions of the tropics. When left without further interference, secondary forests undergo a very slow transition towards climax forest of the original composition.

The successional processes by which a newly opened area develops into climax forest occur in several stages which are well-marked as to their floristic composition. In Central America, Budowski (1965) proposed four stages of tropical forest succession: pioneer stage, early secondary stage, late secondary stage and climax. The species which participate in the first two stages have a fairly wide distribution and within a particular tropical forest they keep re-appearing in large numbers. In section 4.4 some of the genera and species found in different regions are given. Late secondary stage species attain a considerable size and in Africa at least are often recruited from forest formations of relatively drier districts than the regenerating forest itself (see also Taylor, 1960).

Finally, in the climax stage, a balanced community of species is achieved, in which shade-tolerant plants grow in equilibrium with more light-demanding emergent trees. The precise composition depends on climate, soil and position in the catena, and is regarded as a climatic climax, except where extreme soil conditions, such as impeded or excessively free drainage, or gross lack of nutrients, make one prefer to speak of an edaphic climax.

Several difficulties arise in connection with the idea of climax stages in the tropical forest, the most troublesome of which concerns the comparative rarity or even complete absence of seedlings and saplings of many of the species which form the upper tree layer. These species are usually pronounced light-demanders and thus their successful establishment in the dimness of the undergrowth would not be expected. Yet, in West Africa for instance, they may not appear even in the light phase of the smaller gaps which are quickly covered by seedlings of many other species. Published studies on size-class distribution in tropical forests and on processes of natural regeneration are rare. Results of one very comprehensive investigation in a forest in

Southern Nigeria show that the shade-tolerant species of the middle and lower layer decreased logarithmically in numbers as the girth-class increased (Jones, 1955, 1956). With light-demanding emergents, the position is different, for in almost every species there is a more or less marked deficiency of some of the middle size-classes, and sometimes even a total absence of small sizes (see Fig. 4.8). Naturally such size-class distribution must in time be expected to result in marked floristic changes in the horizontal structure.

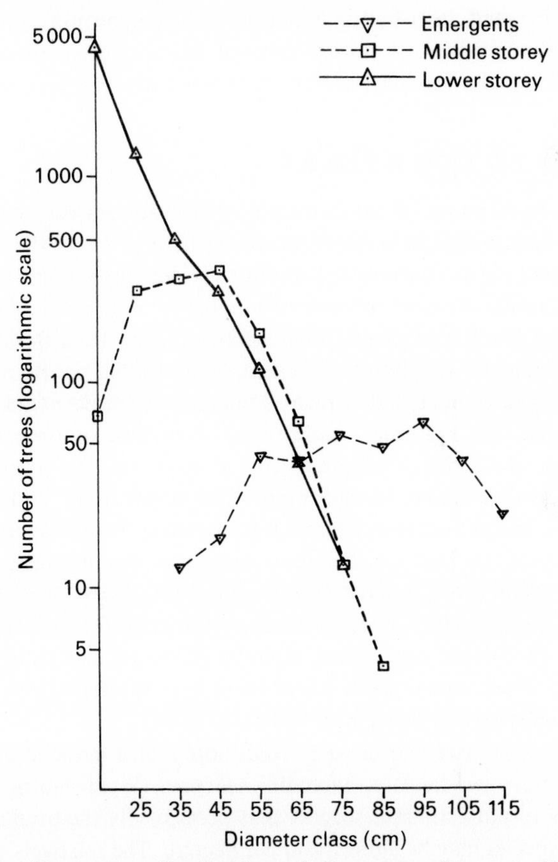

Fig. 4.8 Frequency distribution of diameters of trees in three layers of a Nigerian tropical rain forest. (After Jones, 1956.)

In order to explain this variation of the tropical forest on homogeneous sites, Aubréville's Mosaic or 'Cyclical' theory of regeneration is often quoted (Aubréville, 1938). The composition of the ombrophilous forest, particularly that of the emergent species, is thought to vary both in space and time. There are so many species that the equilibrium of the tropical forest climax can be achieved by different combinations of the available trees. Thus the forest tends to be composed of mosaics of small patches, the composition of which occurs largely at random. The reproduction of any particular patch of forest does not necessarily take place on the same spot, but may occur within other parts of the mosaic. The site of regeneration can be affected by such factors as the production of crops of viable seeds, the influence of seed distributing agencies and competition among the forest species involved. A similar concept has also been discussed by Steenis (1958) who has suggested the term 'spot-wise regeneration'.

4.4 Why so rich a flora?

It is clear from much of the foregoing that there are very numerous life-forms and species in tropical forests, at least as far as the climatic climax vegetation is concerned. Perhaps they are rivalled only by some coral reef communities in richness of numbers and diversity of species.

The total number of plant species given in many local floras is still far from complete. Epiphytes, for example, are rather poorly recorded and it has been estimated that many hundreds of species remain to be described, as, for example, in the case of orchids in New Guinea. Considering the animal kingdom for a moment, the number of unnamed insects alone is enormous. The larger trees and woody climbers are better documented, and it is of course these phanerophytes which provide the bulk of the forest structure. For instance, in five types of tropical forest in Zaïre (Congo, Kinshasa), between 90 and 100 per cent of all plants were phanerophytes, the few others being geophytes (4–10 per cent), chamaephytes (2–4 per cent) and hydrophytes (2–3 per cent), while there were no representatives of the hemicryptophytes and therophytes (Évrard, 1968).

In South-East Asia it is usual to find more than a hundred different species of trees per hectare, excluding seedlings (Wyatt-Smith, 1953), while some recent estimates suggest that occasionally the total number of woody species may be almost 400 per hectare. The relatively poorest Tropical-forest region is the African, where less than 100 woody species per hectare is typical. Even to enumerate the trees over a whole hectare is very laborious, and if the many other vascular plants or indeed the bryophytes are to be included the task becomes even bigger. Yet because of the patchy character of the horizontal structure the minimum area which is theoretically desirable is about 4 ha. Samples of

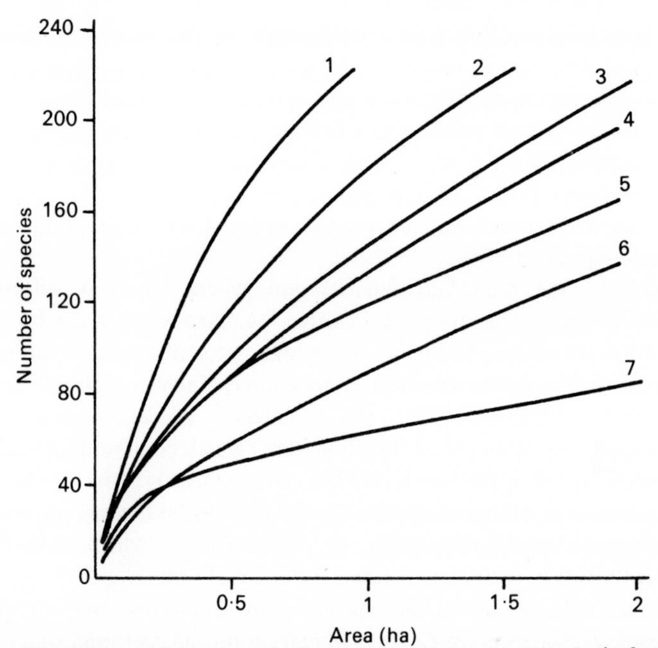

Fig. 4.9 Species/area curves for tree species in tropical forests: 1 – Rengam Forest Reserve, Malaya; 2 – Bukit Lagong F.R., Malaya; 3 – Andulou F.R., Borneo (valley bottom); 4 – Andulou F.R. (ridge); 5 – Mapane, Surinam; 6 – K. Belalong, Borneo (lower hillside); 7 – K. Belalong (ridge). Curves 1, 2 and 5 are for trees exceeding 10 cm diameter; nos. 3, 4, 6 and 7 for those over 9.7cm diameter. (After Ashton, 1964.)

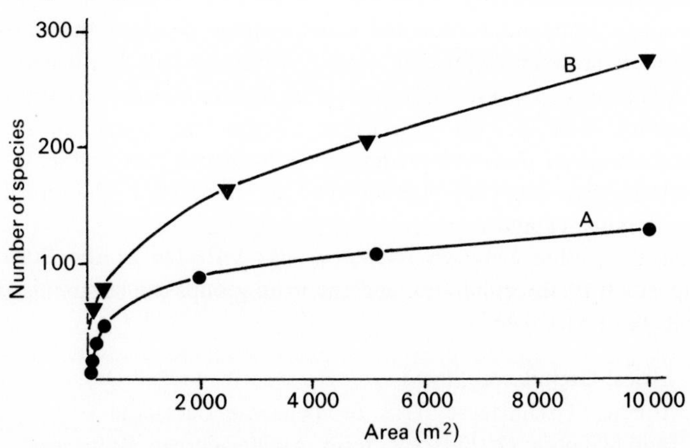

Fig. 4.10 Comparison between species/area curves for (A) tree species and (B) total vascular plants in an African evergreen seasonal forest. (After Lawson *et al.*, 1970.)

this size, however, are hardly manageable in the course of normal ecological studies, and one generally has to rely on extrapolation from smaller enumerations. Figure 4.9 gives a selection of species/area curves for trees in various parts of the tropics, indicating how unreliable a small sample plot can be; it also shows the especial richness of the Indo-Malaysian forests. A rare attempt to compare the tree diversity with that of all vascular plants is shown in Fig. 4.10, based on the work of Lawson *et al.* (1970).

Some of the important families and genera found in different tropical forest regions are given in Table 4.4, from which it can be seen that some are pantropical while other taxa occur mainly or entirely in one part of the world. American, African and Malayan tropical forests are often characterised by the frequency of leguminous trees (Caesalpinaceae, Mimosaceae, Papilionaceae) which commonly appear as emergents in the upper tree layer. The same synusia is often dominated by dipterocarps (Dipterocarpaceae) in the Indo-Malaysian region, where the presence of families such as the Fagaceae, which have tree representatives in temperate forests, is worthy of note.

At the level of genera and species, only a few taxa are naturally pantropical. For instance *Ceiba pentandra* is nowadays found widely in America, Africa and South-East Asia, but is only certainly indigenous to the first region (see Baker, 1965). In many other cases, further taxonomical and phytogeographical studies will be needed to establish the status of closely related species, as for instance *Symphonia gabonensis* in Africa and *S. globulifera* in America (for further examples see Schnell, 1961).

In single-dominant forests and other edaphic climaxes the floristic differences between regions are more conspicuous still. For instance in the American region the leguminous trees *Eperua falcata* and *Mora* spp. frequently form a high proportion of the tree layers. In Africa *Gilbertiodendron dewevrei* predominates in certain parts, while in the Indo-Malaysian countries *Agathis* spp. are sometimes widespread in single-dominant stands.

The distinction between regions is also reflected in the floristic composition of the epiphytes, and the main groups can be summarised as follows (Tixier, 1966):

Africa: Orchidaceae, ferns
America: Orchidaceae, ferns, Bromeliaceae, Cactaceae
Indo-Malaysia: Orchidaceae, ferns, Asclepiadaceae, Rubiaceae

Table 4.4. Examples of families and genera containing dominant, abundant or subendemic species of woody plants in the chief Tropical-forest Regions.

	Family	*Genus*
American Tropical-forest Region	Leguminosae	*Andira, Apuleia, Dalbergia, Hymenolobium, Mora*
	Sapotaceae	*Manilkara, Pradosia*
	Meliaceae	*Cedrela, Swietenia*
	Euphorbiaceae	*Hevea*
	Myristicaceae	*Virola*
	Moraceae	*Cecropia*
	Lecythidaceae	*Bertholletia*
African Tropical-forest Region	Leguminosae	*Albizzia, Brachystegia, Cynometra, Dialium, Erythrophleum, Gilbertiodendron*
	Sterculiaceae	*Cola, Nesogordonia, Tarrietia, Triplochiton*
	Meliaceae	*Carapa, Entandrophragma, Khaya, Trichilia*
	Euphorbiaceae	*Drypetes, Macaranga, Ricinodendron, Uapaca*
	Moraceae	*Antiaris, Chlorophora, Ficus, Musanga*
	Sapotaceae	*Afrosersalisia, Chrysophyllum*
	Ulmaceae	*Celtis*
Indo-Malaysian Tropical-forest Region	Dipterocarpaceae	*Dipterocarpus, Dryobalanops, Hopea, Shorea, Parashorea*
	Moraceae	*Artocarpus, Ficus*
	Anacardiaceae	*Mangifera*
	Actinidiaceae	*Actinidia*
	Daphniphyllaceae	*Daphniphyllum*
	Dilleniaceae	*Dillenia*
	Gonystylaceae	*Gonystylus*
Australasian Tropical-forest Region	Myrtaceae	*Eucalyptus, Agonis, Baeckea, Backhousia, Osbornia*
	Dipterocarpaceae	*Dipterocarpus*
	Casuarinaceae	*Casuarina, Gymnostoma*
	Himantandraceae	*Galbulimima*
	Corynocarpaceae	*Corynocarpus*
	Dilleniaceae	*Hibbertia*
	Menispermaceae	*Carronia*
	Cunoniaceae	*Ceratopetalum*

Floristic differences also appear in the composition of secondary forest. Some representative genera and species are:

Africa: *Harungana madagascariensis, Macaranga* spp., *Musanga cecropioides, Trema guineensis*
America: *Cecropia* spp., *Miconia* spp., *Vismia guianensis*
Indo-Malaysia: *Elaeocarpus* spp., *Glochidion* spp., *Macaranga* spp., *Mallotus* sp.

Considerable changes in species composition have occurred due to the activities of man. The effects of widespread farming have already been discussed, but another potent influence on the local floras has been the selective removal of valuable timber species such as the mahoganies, *Swietenia macrophylla*, *Khaya ivorensis* and *Entandrophragma* spp. This has led to what might be termed 'depleted forests', in which the younger stages of the economic species are often scarce also. A more drastic change in local floras has occurred with the establishment in tropical forest areas of plantations of economically useful 'exotics', such as rubber, oil-palm, teak, *Eucalyptus* spp. and other quick-growing timber trees. The significance of these trends for tropical botany will be further discussed in Chapter 6.

Returning to the untouched primary forest, there is much argument over the possible reasons for its great diversity of species (Corner, 1954; Fedorov, 1966; Margalef, 1968; Richards, 1969). There is no one complete theory, but the salient points appear to be as follows:

1. Any explanation of the present contrasts in species diversity between tropical forests and other vegetation must take into account the underlying environments in which they have developed. Naturally, the more variable the conditions, the more varied the flora. One cannot ascribe the species diversity to 'favourable and unchanging warm and moist conditions occurring over large areas', because such uniformity of environment would necessarily tend to produce a uniform vegetation. In any case, as we have seen, the climate and soils are in fact rather variable, and most strikingly, the structure of the tropical forest itself creates a wide range of microhabitats or niches many of which differ greatly from each other.

2. Another important aspect in the development of a diversified flora is the period of time available for its creation. Other things being equal, the longer the time, the greater the opportunity for the development of new forms and for speciation. Although there have been small-scale paleoclimatic fluctuations (see section 2.2), the greater part of the tropical forest region has presumably not undergone the widespread

and devastating changes of climate which occurred in the north temperate zone. Here entire forest areas were wiped out repeatedly by drastic climatic changes and glaciation in the late Tertiary and Pleistocene, and thus to-day's flora is relatively young and very depleted. Trees, of course, have a long life-cycle and it may well be that a time-scale of 100 000 years or more is needed for the full development of homeostasis in mixed forests.

3. The diversity of a certain area depends, additionally, on the possibility of immigration, with exchange of forms and species. The existence of neighbouring regions of different character, such as mountains and semi-arid regions, will affect the chance of interchange. This will also depend on such factors as the direction of winds and sea currents, the orientation of main river valleys and mountain ranges, and migration routes of birds and herds of mammals. For example, one may expect the floras of the Amazon Basin to be influenced by their accessibility from the neighbouring Andes.

4. The effectiveness of immigration also depends on whether the exchanging forms and species can permanently establish themselves within their new environment. The question here is whether the tropical forests possess special advantages as far as preservation of species goes. Since it has been shown in point 1 above that there are many different niches within the tropical forest structure, it seems clear that this kind of vegetation can indeed serve as a refuge for species. It may be objected that there is severe competition inside the tropical forest, and indeed there is at any one site. But, throughout the mosaic of the forest at large, new gaps are always appearing, providing opportunities for colonisation by both old and invading forms.

5. Diversification also comes about through speciation *in situ.* Is the tropical forest particularly suitable for the development of new forms? In order to answer this, one has to consider the relevant population patterns. Unlike the conditions in temperate forests and many other widely-studied ecosystems, individual specimens or small groups of a species are often isolated from others of their kind. Pollination patterns therefore are quite different from those in communities containing large numbers of each species, and cross-pollination, so important in evolution, may be restricted in extent except where specialised adaptations are in play, as, for example, gregarious flowering (see section 5.8). The general isolation, however, will tend to produce new forms through genetic drift. Thus internal isolation occurs within the same vegetation structure that is potentially accessible to invading species. In addition speciation may be enhanced through violent but local disturbances, such as hurricanes, volcanic eruptions and mountain building.

6. All in all the reasons for species diversity in tropical forests are very complex, showing many interactions of the environmental and evolutionary features which have been discussed above. Moreover, one should be ready to consider that there may be special biochemical or biophysical characteristics which favour both luxuriance and speciation in the warm and humid tropics. An obvious example would be if mutations occurred more frequently in tropical forest, for it is well known that higher temperatures promote mutation rates of organisms under experimental conditions. A number of high-yielding tropical plants, such as maize and sugar cane, have been shown recently to differ from many other crops in lacking photo-respiration, and thus in conserving carbohydrates. It remains to be seen how widespread this phenomenon is among other tropical plants, but it could possibly be a factor in the high primary production and the general luxuriance of the tropical forest. These might also be influenced by biophysical considerations, for instance by changes which occur in the properties of water molecules at relatively higher temperatures, or by other biochemical aspects, such as generally increased enzyme activity occurring under the prevailing mild temperatures in the region of 25°C.

4.5 Attempts at classification

As with other vegetation types, a reasonable classification of tropical forest is one of the desired aims in ecological research. Most definitions of vegetation units and synsystematic studies have been limited to a particular country, island or other small area. Only a few of them have a world-wide coverage, such as the classifications by Burtt Davy (1938) and Ellenberg and Mueller-Dombois (1967). Lack of pantropical species, difficulties of size and accessibility, and the overall complexity of tropical forests do not allow a very detailed floristic and ecological description, as is usual, for example, in the Zürich-Montpellier School. The physiognomic characters of the vegetation structure, combined with the broad features of the pertinent relief and soil, appear to provide the only possible basis for a world-wide classification.

On a more local scale, the methods are too variable to allow a satisfactory summary to be given. The older 'physiognomic' approach does not satisfy the usual requirements of forestry and land-use, and phytosociological methods have more frequently been applied in order to distinguish more detailed and economically manageable subdivisions (see Lebrun, 1947; Schnell, 1952; Lebrun and Gilbert, 1954; Germain and Évrard, 1956; Schmitz, 1963; etc.). The basic vegetation unit is the 'association', which is further arranged in a hierarchy of higher units, namely 'alliance', 'order' and 'class'. It is standard practice to give the vegetation units single or double names composed of the Latin names of the component species. This terminates in a specific ending which indicates the unit's rank. In Katanga, Zaïre (Congo, Kinshasa), the closed tropical forests were classified by Schmitz (1963) into two classes, Mitragynetea and Strombosio-Parinarietea, the latter being subdivided as follows:

Class: Strombosio-Parinarietea
- *Order*: Piptadeniastro-Celtidetalia
  - *Alliance*: Albizzio-Chrysophyllion
    - *Association*: Mellereto-Canarietum
    - *Association*: Klainedoxeto-Pterygotetum
  - *Alliance*: Albizzio-Chrysophyllion
    - *Association*: Entandrophragmeto-Diospyretum hoyleanae
    - *Association*: Alchorneeto-Voacangetum africanae

Many authors, however, deny that floristic principles can be used in the classification of tropical forests, or that the concept of the 'association' fits these conditions. The analytic phase of the phytosociological method certainly cannot be used without adequate alteration of the field procedures developed in temperate countries. In many cases classifications according to soil conditions and catenas will provide a satisfactory background for surveying, land-use and forestry management.

The following is the tentative physiognomic–ecological classification of tropical forests, based with minor alterations on the work by Ellenberg and Mueller-Dombois (1967):

A. *Tropical ombrophilous forests.* (Tropical rain forests in the strict sense.) Composed mostly of evergreen trees with little or no bud protection. Not particularly cold- or drought-resistant. Truly evergreen forest, i.e. some individual trees may be leaf-exchanging, but not simultaneously with all the others. Many species possess leaves with drip-tips, at least when young.

1. Tropical ombrophilous lowland forest. Multilayered structure composed of several tree layers and discontinuous ground-herb storey. In the upper tree layer many trees exceed 30 m height and form large buttresses. Numerous species are associated in the closed middle tree layer. Sparse undergrowth is composed mainly of regenreating trees. Palms and other unbranched trees rare, woody lianes nearly absent. Vascular epiphytes less abundant than in A2 and A3.

2. Tropical ombrophilous mountain forest. Tree sizes markedly reduced, few species exceeding 30 m height. Tree crowns extending farther down the stem than in A1. Undergrowth abundant, often represented by tree-ferns or small palms. Abundant vascular and other epiphytes. The ground layer rich in herbs and bryophytes. Lianes present. (Corresponds most closely to the 'textbook' description of the 'virgin' tropical rain forest.)
   a. Broad-leaved, the most common form.
   b. Needle-leaved or microphyllous, coniferous forest.
   c. Bamboo, rich in tree-grasses replacing largely the tree-ferns, pigmy trees and palms.
3. Tropical ombrophilous cloud forest. Closed forest with numerous gaps and liane thickets. Trees often gnarled, rarely exceeding 20 m in height. Tree crowns, branches, limbs and trunks heavily burdened with epiphytes. Numerous woody lianes and herbaceous climbers. The ground extensively covered by mosses, liverworts, herbaceous ferns, *Selaginella* spp. and broad-leaved vascular herbs.
   a. Broad-leaved, the most common form.
   b. Needle-leaved, or microphyllous, coniferous forest.
4. Tropical ombrophilous alluvial forest. Multilayered closed forest with frequent herbaceous undergrowth, palms common, and more frequent vascular epiphytes than in A1. Trees often with buttresses and stilt roots. Numerous gaps caused by the shorter life-span of emergent trees.
   a. Riparian. Narrow strip along the river bank at the lowest level that supports forest, frequently flooded. Limited number of tree species, often much-branched. Shrubs resistant to river action. Herbaceous undergrowth nearly absent, epiphytes very rare.
   b. Occasionally flooded. On relatively dry terraces along active rivers, most extensive of the A4 type. Some emergent trees achieve giant sizes of over 50 m height, with buttresses reaching 10 m height. More epiphytes than in A4(a) or A4(c). Numerous woody and herbaceous climbers.
   c. Seasonally waterlogged. Along river courses with the water stagnating for several months, sometimes twice or several times during a year. Upper tree layer rather broken and uneven in height. Trees often with stilt-roots. Middle and lower tree layer often dominated by a few species capable of vegetative reproduction. In open places palms and taller herbs occur.
5. Tropical ombrophilous swamp forest. In depressions and smaller

valley bottoms with more or less permanent excessive soil moisture and poor soil aeration. Similar to A4(c), but as a rule poorer in tree species and with extensive mats of ferns or grasses forming the undergrowth. Trees less than 30 m high, frequently with stilt-roots, buttresses and pneumorhizae, such as peg-roots and knee-roots.

a. Broad-leaved, dominated by dicotyledonous species.
b. Dominated by palms, but broad-leaved trees scattered in the undergrowth.

6. Tropical evergreen peat forest. On nutrient-poor soils covered by an accumulating organic layer. Stands lower than 20 m height except in some South-East Asian islands, composed of only a few, slow-growing, broad-leaved trees or palms. Trees are commonly equipped with pneumorhizae and stilt-roots. Few ground-herbs, mostly represented by ferns.
   a. Broad-leaved, dominated by dicotyledonous plants.
   b. Dominated by palms, forming pneumorhizae and pneumathodes.

B. *Tropical or subtropical evergreen seasonal forests.* Mainly evergreen and leaf-exchanging trees with some bud protection. Tall species (up to 40 m) of the upper tree layer are moderately drought-resistant and shed leaves particularly during the dry season though not all simulaneously. Transitional forest between A and C.

1. Tropical or subtropical evergreen seasonal lowland forest, the commonest.
2. Tropical or subtropical evergreen seasonal montane forest. In contrast to A2 no tree-ferns are present. Evergreen shrubs and numerous climbers in the undergrowth.
3. Tropical or subtropical evergreen dry subalpine forest. Physiognomically resembling Mediterranean sclerophyllous forests. Dense stands of trees with sclerophyllous leaves, achieving a height of about 10 m. On the dark forest floor there are no ground-herbs or shrubs. Epiphytes represented mostly by lichens.

C. *Tropical or subtropical semi-deciduous forests.* The majority of the emergent trees in the upper tree layer are quite drought-resistant and shed leaves fairly regularly during the dry period. Some of the emergents achieve a height of 40 m. Middle tree layer, pigmy trees and shrubs in the undergrowth are evergreen and may have sclerophyllous leaves. The majority of trees with some kind of bud protection, leaves without drip-tips.

1. Tropical or subtropical semi-deciduous lowland forest. Multi-layered structure with marked emergents in the upper tree layer. Some of these achieve a height of 30 m. In the undergrowth mostly tree seedlings, tree saplings and woody shrubs. Practically no epiphytes present, but both annual and perennial herbaceous lianes represented. Among the ground-herbs many graminoid species.
2. Tropical or subtropical semi-deciduous mountain or cloud forest. Structures similar to C1, but trees not so tall, and covered with xerophytic epiphytes (e.g. *Tillandsia usneoides*).

D. *Subtropical ombrophilous forest.* Stands locally developed in Northern Australia and on Taiwan, with more pronounced temperature differences between summer and winter. Multilayered stands are similar to the tropical stands, but trees are less vigorous and allow more shrubs to grow in the understorey. The subdivisions can conform to those under A.

E. *Mangrove forests.* Halophilous forests in the intertidal parts of the tropical and subtropical zone. Trees growing almost entirely in a single layer, achieving a height of up to 30 m. Trees equipped with evergreen sclerophyllous leaves and various kinds of stilt-roots, peg-roots, knee-roots and pneumathodes. Epiphytes are generally rare except for a few algae, bryophytes and lichens attached to aerial roots and lower parts of stems. Ground herbs only exceptionally present (e.g. the fern *Acrostichum aureum*).

Chapter 5

The physiology of tree growth

In the previous chapter, tropical forests and the trees in them were considered mostly as static structures, although it was emphasised that the forest is continually changing through growth and reproduction, death and germination. These more dynamic, physiological aspects will now be taken up in more detail, and the growth of the tropical tree examined against the background of the climatic and other factors discussed in Chapter 3. Other branches of plant physiology, such as metabolism, nutrition and translocation, will be mentioned only where they have particular relevance to the central problem of the growth of the whole tree.

Growth implies the self-multiplication of living material, and so in its widest sense includes all increases and changes which occur in the life of a plant. A convenient distinction is often made, however, between quantitative increase, or growth in its narrower sense, and development, which covers definite changes in the production of various plant organs and tissues. For example, measurements of the rate of expansion of a leaf blade or elongation of its petiole come into the first category; while assessments or observations on the emergence of new leaves from the bud, the duration of their growing period and their ultimate loss from the tree come into the second.

The size and shape attained by the mature leaf is thus governed both by its growth and its development. Variation between leaves occurs, even on a single tree, because the same genetic potential has been expressed in environments which are different, and also because of controlling systems within the tree. These internal systems are quite difficult to study (see section 5.7), and in any case require some understanding of the effects of environment, and it is therefore logical to start a physiological investigation by considering the various external factors which may affect growth and development. This can be done, for instance, by growing potted tree seedlings under conditions which are uniform except for the factor or factors to be studied.

Experiments of this kind in controlled environments can thus provide information on the responses of the species to individual factors of the environment. A sound, scientific basis can then gradually be built up from which to predict the likely effects of climatic and other factors on tree growth in the tropical forest. In time it should be possible to identify the key features or changes in the forest environment which 'trigger off' such developmental processes as the initiation of flower or root primordia, the outgrowth of new shoots from the bud or the germination of seeds. Other environmental variations may prove to be important in a rather different way, by making the conditions more favourable or less suitable for growth to take place, but not acting as 'signals' to the plant in the same way. A similar distinction has been made between 'proximate' and 'ultimate' factors influencing periodicity in tropical animals (Owen, 1966).

As has been mentioned in Chapter 1, it is often imagined that all the trees in the tropical forest are producing new leaves and stems throughout the year. That shoot growth is typically periodic, with intervals of activity and rest alternating with each other, was stressed many years ago by Schimper (1898) and Coster (1923), and has been reiterated by Richards (1952) and recently Zimmermann and Brown (1971), but many authors still refer to 'continuous growth in the tropics'. On the other hand, Kramer and Kozlowski (1960) go too far in stating that 'no matter how favorable the environment, woody plants do not grow continuously but alternate between flushes or periods of activity and periods of inactivity or dormancy'. The proportion of woody species found to grow continuously after the seedling stage was less than 20 per cent in Java and Malaya (Coster, 1923; Koriba, 1958), and many of these were shrubs and small trees. It may well be that the only groups of large forest plants in which continuous growth is common in the natural habitat are the palms, conifers and tree-ferns.

Once periodicity of shoot growth has been accepted for the majority of the dominant members of the forest community, it becomes very important to know whether their periods of growth and rest are to any extent synchronised or seasonal, or whether they are completely random. The latter view was taken by Warming (1909), who stated that 'there is no periodicity in the life of the forest as a whole', but the evidence suggests that this is not so for tropical evergreen seasonal forests, and it is probably incorrect also for ombrophilous forest even when the climate varies only slightly (see Fig. 2.2). Seasonal peaks of flowering, shoot growth or leaf-fall can be masked by the presence of large numbers of different species, and by the considerable variation between the behaviour of individual trees or even parts of a single tree. An area of more or less natural forest has to be studied at frequent intervals for a number of years before the events which occur regularly can be detected.

The main aspects of growth and development of tropical forest trees are covered in the following sections, and information is given on seasonal, non-seasonal and irregular patterns. Experimental evidence on the effects of environmental factors is stressed, and an attempt is made to correlate physiology and ecology by considering the possible environmental control of tree growth in the forest setting.

5.1 Bud-break

The commencement of a period of shoot growth after an inactive phase is generally first observed as an elongation of minute leaves, though this may be preceded by the enlargement of the bud-scales where these are present (see section 4.2). The subsequent rapid expansion of the leaves and the elongation of the internodes between them is referred to as flushing, and in some species is quite a striking feature of the forest, often described in the literature. However, the important physiological changes leading to renewed shoot growth, must of course take place some time before the first visible signs of outgrowth can be detected.

The impression gained by Simon (1914) at Bogor in Western Java, where the climate is fairly constant, was that there was probably no month without flushing. This may well be true of more seasonal climates too, but it is perhaps more a reflection of the variability of the forest than its uniformity. Many authors have shown that there is increased flushing at certain times of year, and less at other times (e.g. Coster, 1923; Taylor, 1960; Njoku, 1963; Hopkins, 1970). Some trees appear to flush regularly, but to be 'out of step' with the calendar months, particularly in a rather uniform climate such as that of Singapore (Holttum, 1940). Others are irregular in flushing, and there is usually also a good deal of variation between trees of the same species. Even so, a tendency to seasonal flushing can still be detected in many trees, and appears to be an important feature of many tropical forests.

Alvim (1964) has drawn attention to the frequency with which peaks of flushing occur at or near the equinoxes (see Fig. 5.1), and this appears to hold true both in the northern and southern hemispheres, and even very close to the Equator. Certain trees, such as *Terminalia catappa*, *Peltophorum pterocarpum* and *Cola acuminata* frequently flush twice a year, during both equinoctial periods. Others, such as cocoa, for example, flush even more often when unshaded, but tend to exhibit major peaks around the equinoxes when grown under shade

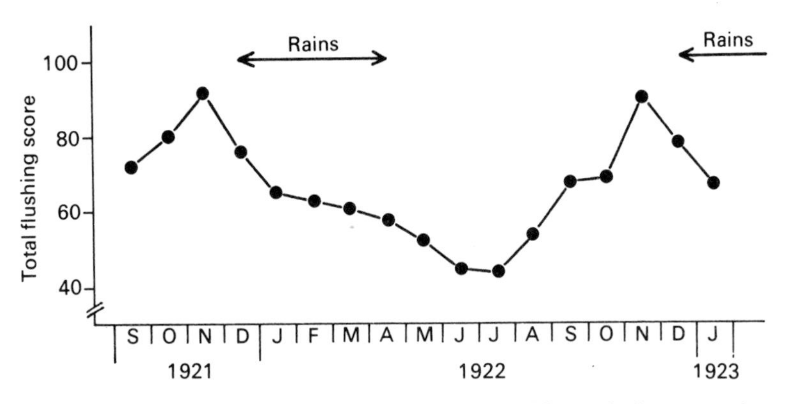

Fig. 5.1 Frequency of flushing by months in 52 tropical tree species at Toeban, W. Java. Scored on a flushing scale of 0 – 3. (Compiled from data of Coster, 1923.)

conditions, more comparable with their natural environment (Fig. 5.2). Although these studies do not all refer to indigenous plants growing in natural or semi-natural forest, they are of interest in that quite large numbers of trees have been observed, growing in the Indo-Malaysian and African tropical forest regions.

Bud-break can therefore occur at any time in the year, but is more likely at certain seasons. It is a remarkable fact that in evergreen seasonal forest, flushing frequently occurs in the dry season, with the new shoots emerging some time before the rains start in earnest. At first sight this seems very surprising, since the rainy season is so obviously the growing season for many crop plants, grasses and other herbaceous species. The same does not necessarily apply to trees in the forest, however, and it is clearly very important to clarify when the growing season for the majority of woody species occurs, and against what ecological background. Some indication will be given in section 5.3 of possible 'ultimate' factors which may have influenced the evolution of such seasonal flushing (see also Table 5.3), but attention will now be focused on the physiological factors which may be at work in 'triggering-off' the bud-break.

It is often tacitly assumed that flushing is caused by rainfall, but this cannot of course be true for the many species commencing growth during dry weather. Even for those which flush after rain there is usually no more than circumstantial evidence that water is directly involved. Njoku (1964) has issued a timely warning about the dangers of trying to fit crop growth measurements to rainfall data and infer a causal relationship. Correlation alone, however many complex statis-

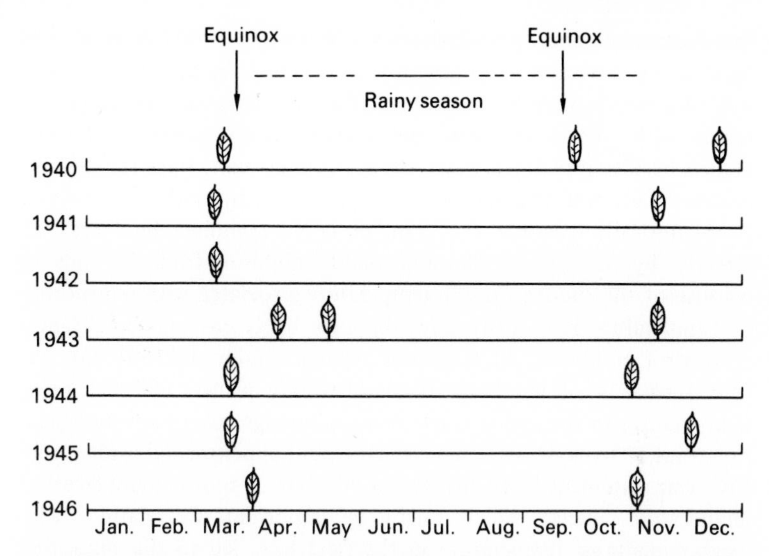

Fig. 5.2 Occurrence of flushing peaks in cocoa plantations grown under shade at Tafo, Ghana (6°N). The symbols represent the times when half or more of the trees were flushing. The time of bud-break is estimated to be about two weeks before the flushing peaks. (Compiled from data of Greenwood and Posnette, 1949.)

tical tests show significance, can provide only hypotheses, and experimental methods are essential to avoid making serious errors.

It is possible that in teak an increase in the water status of the shoot may cause bud-break. Coster (1923) found that simply standing cut leafless shoots in pure water was usually enough to bring about bud-break in 7–10 days, whether they were inside a room or out-of-doors. Similar effects were found in preliminary studies by Njoku (1963) with *Terminalia superba*, *Bosqueia angolensis* and *Millettia thonningii*. The most likely explanation of these results is that the buds were in a state of 'post-dormancy' (Wareing, 1969). Thus they were probably quiescent at the time, requiring only an increased supply of water for bud-break, rather as a seed may lack only water for germination.

However, the same authors found that cut twigs of *Bombax malabaricum*, and *Monodora tenuifolia* and *Sterculia tragacantha* could only be forced in water when taken either early or late in the resting period. This suggests that these trees are similar to the many temperate woody plants showing a period of 'true dormancy', when the buds will not grow under the conditions which are favourable for shoot extension growth. The early and late periods, when outgrowth of shoots may be possible under favourable conditions, are referred to respectively as 'pre-dormancy' and 'post-dormancy' (Wareing, 1969). A great deal more study is needed before the role of water can be properly assessed, but two points may be made here. Firstly, observations and experiments with potted seedlings and grafted plants suggest that other factors besides prolonged water stress are important, since these plants received daily watering and yet some grew intermittently. Secondly, it is theoretically possible that rainfall could stimulate bud-break indirectly, by the leaching of water-soluble inhibitors from the buds, or because of the sudden drop in temperature associated with rain-storms.

Temperature is in many ways the most likely environmental factor to control bud-break. In temperate regions, winter chilling breaks the true dormancy of the buds of many woody plants, and then rising temperatures in the spring allow them to emerge from post-dormancy. For tropical trees, there is not yet sufficient experimental evidence for any clear statement to be made, and there are of course many facets of temperature change which might be important. Correlations with measurements of temperature in the field have led to the suggestion that cocoa buds may tend to flush when the maximum temperatures exceed about 28°C, and when the daily range of temperature exceeds 9°C (Hardy, 1958, 1964; Alvim, 1967; see also section 3.2). Experiments with 'mature-wood' cuttings of cocoa grown in growth-rooms were carried out by Murray and Sale (1966, 1967), who found much more frequent flushing at 30° or 31°C than at 23°C. Day temperatures appeared to be rather more important than night, but there was no suggestion that daily fluctuation of temperature promoted flushing of this species.

Day-length treatments can be used to prevent or stimulate bud-break in some temperate woody plants, particularly those which do not have a chilling requirement, and it is of considerable interest to know whether tropical trees are influenced at all by this factor. Alvim (1964) suggested that cocoa grown at latitudes 10–15° showed a period without flushing because the days were shorter at that time. However, he was not able (Alvim and Grangier, 1965) to stimulate flushing in a plantation by increasing the day-length to 15–16 h with bright lights, and he concluded that bud-break in cocoa was not after all photoperiodically controlled.

Experiments in growth-rooms have shown that bud-break in two tropical trees is clearly influenced by day-length (Longman, 1969). Leafy seedlings of the West African species *Hildegardia barteri* remained dormant and did not flush under short-days (9 h 10 m). When most or all the leaves had become yellow or fallen off, or where the seedlings were already leafless at the beginning of the experiment, bud-break occurred. Under long-days (17 h 10 m), flushing took place whether leaves were present or not. With *Cedrela odorata*, originating from Guyana, S. America, bud-break in 'mature-wood' grafts was also prevented by short-days until the leaves were lost from the plants. The buds in these experiments appear to have been pre-dormant; that is to say prevented from flushing through inhibition by mature leaves under short-days (see Wareing, 1969). Release from this type of dormancy was effected either by loss of the leaves, or by transfer to long-days. It is interesting to note in this connection that out-of-season bud-break can be induced when *Brachystegia laurentii* is defoliated by caterpillars (Germain and Évrard, 1956), while experimental defoliation stimulated flushing in *Couroupita guianensis* grown in a greenhouse at Heidelberg in Germany (Klebs, 1926).

Further work may show to what extent flushing under natural conditions is controlled by these and other climatic factors, but meanwhile it is unnecessary to invoke yearly endogenous rhythms (Bünning, 1948) to account for the regular occurrence of bud-break in the dry season. Various types of internal controlling factor may prove to be important in cases of non-seasonal flushing, in leaf-exchanging species (section 5.5), and in trees where the vegetative buds do not grow out while flowering is occurring, for example *Ceiba pentandra*, *Hildegardia barteri*.

5.2 Rate of shoot elongation

A bud which is flushing actively produces a new stem by the simultaneous formation and extension of internodes, or by the simple elongation of pre-formed tissue. Some trees combine both systems of growth. During the period of active outgrowth the new leaves are also expanding, but for convenience this will be considered in section 5.4 along with other aspects of leaf growth. The process of leaf production will, however, be considered here.

The rate of stem elongation can be extraordinarily high; up to 0·9 m/day has been recorded for a species of bamboo, and over a longer time scale more than 20 m in seven years for *Albizzia moluccana* in the Andaman Islands (Bradley, 1922). A comparison between quickly growing pioneers and climax forest species in Zaïre (Congo, Kinshasa) is presented in Fig. 5.3, based on the work of Lebrun and Gilbert (1954). It is interesting to note that the high growth rates of the pioneers tend to decline somewhat within 5–8 years, whereas the climax species are starting to grow faster. Average rates of early height growth were 2·8 m/year for *Terminalia superba* and 3·8 m/year for *Musanga cecropioides* (compare Plate 6). Faster rates still, reaching 5·5 m/year, have been recorded for plantations of *Ochroma lagopus* (Anonymous, 1960), while a young individual of *Cedrela odorata* in the

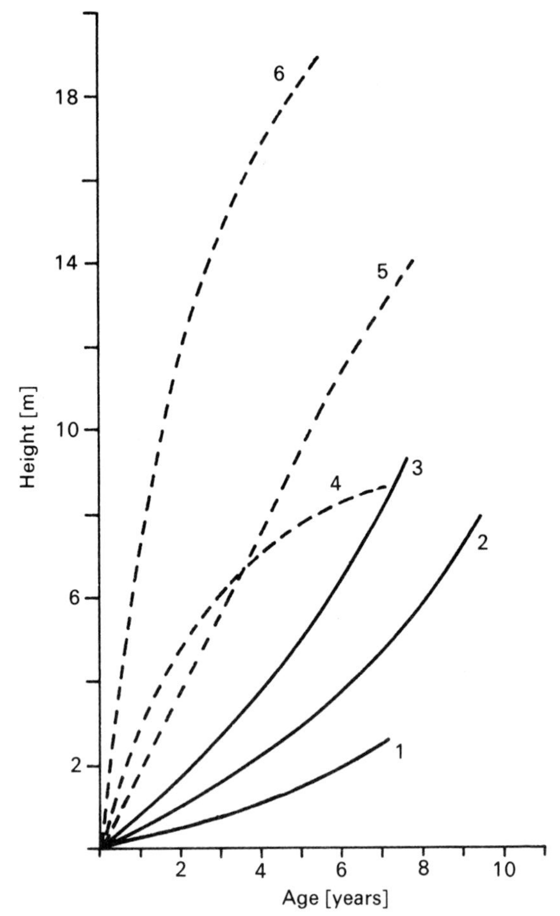

Fig. 5.3 Growth in height of young trees growing under natural conditions in Zaire (Congo, Kinshasa); species of climax forest shown with a full line; pioneer trees in clearings have dotted lines: 1 – *Scorodophloeus zenkeri*; 2 – *Oxystigma oxyphyllum*; 3 – *Gilbertiodendron dewevrei*; 4 – *Calancoba welwitschii*; 5 – *Terminalia superba*; 6 – *Musanga cecropioides.* (After Lebrun and Gilbert, 1954.)

Sapoba Forest Reserve, Nigeria, had grown at an average rate of 6·7 m/year.

On the other hand, it is easy to overemphasise tropical growth rates, which are usually found to be lower within the forest community. Exceptional figures are generally derived from measurements on terminal buds of young trees growing on forest margins or in clearings. The shoots on larger trees in the forest are subject to competition from other parts of the same individual and from other plants. There can also be large differences between species; thus *Scorodophloeus zenkeri* grew only about 0·3 m/year, while *Gilbertiodendron dewevrei* grew four times as fast (see Fig. 5.3). At the top of old trees, growth rates are frequently low, sometimes only a few centimetres in a year.

All the environmental factors discussed in Chapter 3 can influence the rate of stem extension, as well as certain others such as the force of gravity (Damptey, 1964). Perhaps one of the most important is light intensity, because of its controlling effect on the rate of photosynthesis, which in turn influences the other syntheses involved in growth. In the field, however, and even in growth-rooms, the effects of changes in light intensity are peculiarly difficult to assess. Some of the reasons for this have already been discussed in section 3.1, but in addition there is the problem of distinguishing between direct effects of light on photosynthesis, and indirect effects such as higher leaf temperatures, increased water loss, changed stomatal aperture, etc. Further information can be obtained from such sources as Murray and Nichols (1966), Kozlowski and Keller (1966) and Hughes (1966).

The growth of woody plants in relation to water has also been very widely studied (see for instance Kramer and Kozlowski, 1960; Rutter and Whitehead, 1963; Zahner, 1968). It appears to be a feature of many plants that the rate of growth is reduced as soon as quite small water deficits develop, while if the point is reached when the cells of the leaf begin to plasmolyse, growth comes to a complete halt. The reason for this reduction or cessation of growth seems to be that the rate of photosynthesis is similarly affected and the two processes are generally closely linked (Stocker, 1960), partly because increased levels of abscisic acid may lead to stomatal closure.

Daily fluctuations in growth rate, as well as longer-term effects, can result from changes in the water status of the plant. Automatic records of the extension of a young shoot of a tropical bamboo made by Coster (1927) showed that it grew at a rate of 13 mm/h during the night, but there was a very pronounced check to growth during a hot day. The following day was also hot and sunny, and by 11.30 hours the rate had fallen to 5 mm/h. At this point Coster removed the four tall canes growing from the same clump, whereupon the growth rate rose before noon to 16 mm/h. Since the change was so large and so rapid, the conclusion is almost inescapable that growth had been reduced by temporary water stress, and that it increased because the main transpiring surfaces had been removed. Coster also found a species which grew at a rather steady rate throughout a night and a hot day (the climber *Aristolochia gigas*), and a case where day-time growth exceeded that made in the night (*Congea villosa*, a liane). The latter condition was also found by Longman (1964) in a young yam shoot (*Dioscorea* sp.), and so it seems unwise to assume that shoot growth is typically greater at night (see for example Went, 1957).

It has been shown in section 3.2 that there may be considerable air temperature fluctuations in the tropical forest, particularly at the higher levels and in gaps. The seasonal changes are not as great as those often experienced by trees in the temperate zone, and it has sometimes been concluded that there would be only minor effects on tropical tree growth. Evidence is accumulating, however, which suggests that many species may be very sensitive to temperature changes. Growing potted plants at different air temperatures in controlled environment rooms is necessary to demonstrate this, since all of them can then receive similar lighting conditions and adequate water. In this way, for instance, Kwakwa (1964) showed that raising the temperature at which *Ceiba pentandra* seedlings were growing from 15° to 36°C increased the shoot growth rate twenty-three-fold. Similarly, when the night temperature only was raised from 20° to 30°C, main stem growth was greatly stimulated in several tree species, including *Terminalia ivorensis* and *Triplochiton scleroxylon* (Longman, 1966; see Plate 20 and Table 5.1). Cocoa appears to be a particularly sensitive plant, since raising only the day temperature by 3·5°C increased the growth rate by 250 per cent (Murray and Sale, 1966).

Table 5.1. Effect of night temperature on shoot elongation of seedlings of tropical forest trees

Species	*Night temperature (°C)*		*Growth response*	
	lower	*higher*	*Rise in temperature (°C)*	*% increase in shoot growth*
Cedrela odorata	20	30	+10	+ 94***
Triplochiton scleroxylon	20	30	+10	+ 65***
Ceiba pentandra	26	31	+ 5	+ 74***
,, ,,	31	36	+ 5	+ 6 n.s.
Bombax buonopozense	26	31	+ 5	+ 72***
,, ,,	31	36	+ 5	+ 21*
Gmelina arborea	26	31	+ 5	+140***
,, ,,	31	36	+ 5	− 14 n.s.

The seedlings were grown in soil in pots, and were from 1 to 15 months old at the beginning of the experiments. There were between 10 and 14 plants in each treatment, and they were kept out-of-doors or in a lightly-shaded greenhouse during the day, and in the appropriate growth-room at night. The night temperatures were in operation for between 9½ and 13 h, and treatment lasted for 5–21 weeks. (After Longman, 1972.)

*** significant at the 0.1% level
** significant at the 1% level
* significant at the 5% level
n.s. not significant

One important ecological conclusion from these studies is that such plants are likely to show clear responses to temperature changes in the forest of 1°C or even less, though of course they may not be equally sensitive at all parts of the range of temperatures at which they will grow. Little is known, as a matter of fact, about this range, but for *Ceiba pentandra* Kwakwa suggested that the minimum temperature for shoot growth was probably not much below 15°C, as all seedlings receiving this temperature by day or by night made very poor growth (see Fig. 5.4). Went (1962) has pointed out that some tropical plants

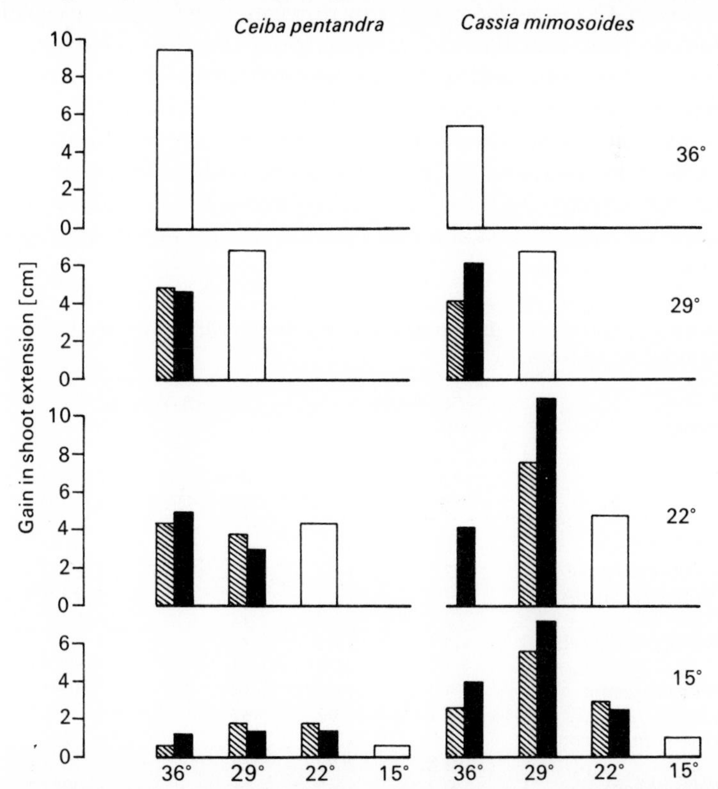

Fig. 5.4 Effect of constant and fluctuating air temperatures on the rate of terminal shoot elongation in young seedlings of *Ceiba pentandra* and *Cassia mimosoides*. Temperature changes effected by moving trolleys of plants to the appropriate constant-temperature growth-room every 12 h; day-length was 13.2 h. The height of the histograms represents the mean gain in height in 10 weeks in *Ceiba* and 6 weeks in *Cassia*. Open columns – constant temperatures; solid columns – fluctuating temperatures, cooler night; shaded columns – fluctuating temperatures, warmer night. Note that the *Cassia* plants receiving 22°C day/36°C night almost all died before the 6th week. (After Kwakwa. 1964.)

may die if kept at 10°C, though the reason for this is not known. Maximum temperatures for shoot growth of *C. pentandra* appear to be over 40°C.

A number of cases have been mentioned in which increasing the temperature greatly stimulated growth (see Table 5.1). In these species it seems that the optimum temperature for shoot elongation is probably quite high. In young seedlings of *C. pentandra* and the shrub *Leucaena leucocephala* it was found to be around 36°C (Kwakwa, 1964), a high value which will rarely be experienced by seedlings in nature. Another feature of *C. pentandra* is that growth seems to be more rapid with constant than with fluctuating temperatures (Fig. 5.4). Moreover, it appears that interchanging the day and night temperatures does not affect the growth rate appreciably. In other species, such as the small woody plant *Cassia mimosoides*, there are different optima for day and for night temperatures. Young coffee plants made most growth with the combination 30°C day/23°C night (Went, 1957), while several temperate region species also make most shoot growth with the nights cooler than the days.

Table 5.2. Effect of day-length on shoot elongation of seedlings of tropical forest trees

Species	*Day-length* (h)		*Growth response*	
	shorter	*longer*	*Increase in day-length*	*% increase in shoot growth*
Cedrela odorata	11·0	14·5	+3½	+ 51**
Theobroma cacao	12·5	14·5	+2	+ 31**
Terminalia superba	9·2	13·2	+4	+195***
,, ,,	13·2	17·2	+4	+ 59**
Chlorophora excelsa	9·2	13·2	+4	+ 22 n.s.
,, ,,	13·2	17·2	+4	+355***
Ceiba pentandra	9·2	13·2	+4	+130***
,, ,,	13·2	17·2	+4	− 1 n.s.
Bombax buonopozense	9·2	13·2	+4	+ 82***
,, ,,	13·2	17·2	+4	− 7 n.s.
Hildegardia barteri	9·2	17·2	+8	+200***

All plants received approximately the same total light energy, the extension of the day-length being given with low intensity illumination (20–200 lux). The maximum day-length 5 degrees from the Equator is approximately 13·2 h, allowing for dawn and dusk. Between 11 and 20 plants were used in each treatment; other details as Table 5.1. (After Longman, 1972.)

Since under natural conditions nights are usually several degrees cooler than days, it is easy to conclude that this will necessarily be reflected in the responses of the plant. It has also been assumed that cooler nights favour growth by reducing respiration and thus conserving photosynthates. However, there are three lines of evidence which suggest that these are misleading generalisations. Firstly, as we have seen, there are species where it is immaterial whether a particular pair of temperatures is arranged with the day cooler or the night cooler. Secondly, there are several trees in which rather warm nights clearly promote shoot growth, and even some species which actually have a higher night temperature optimum than day, for example the tropical herb *Saintpaulia ionantha* (Went, 1957) and the subtropical conifer *Pinus ponderosa* (Callaham, 1962). Thirdly, in some species which do have a lower night optimum than day, it is quite easy to show that if the day temperature is held well below its optimum, the reduction in growth can be partly offset by increasing the night temperature so that it equals or exceeds the day temperature (e.g. *Cassia mimosoides*, Fig. 5.4; tomato, Hussey, 1965).

The increased stem extension in warm nights is usually a combination of longer internodes with a more rapid production of leaves. The latter provides a possible explanation for the promotive effects of warm nights, for the greater leaf area can lead to higher photosynthesis and increased dry weight per plant (Hussey, 1965).

Certain temperate trees, for example many cypresses, grow more slowly under shorter day-lengths than they do under long-days. It has generally been assumed that tropical plants would be insensitive to photoperiod, but at least 14 tree species have now been shown to grow faster under long-days (see Table 5.2, and Plates 19A, 20 and 21). Nor are these effects small: for example in *Terminalia superba* seedlings, the growth rate was trebled by increasing the day-length from just over 9 h to the value found in June in southern Ghana. So sensitive is this species, in fact, that significant effects of day-length could be detected after as little as three days' treatment (Longman, 1966). In seedlings of the emergent species *Chlorophora excelsa*, there was an even larger increase when the photoperiod was extended to about 17 h.

These experiments were conducted in growth-chambers and greenhouses in such a manner that the effects were not due to greater amounts of light energy occurring in the longer days. This was achieved by arranging for all treatments to receive approximately the same time under daylight or bright artificial lights, and to extend the day-length with supplementary illumination too dim to have a significant effect on photosynthesis. It is therefore clearly established that many tropical trees are sensitive to photoperiod.

More leaves are produced under longer days in the two species just mentioned, and in *Hildegardia barteri*, coffee, cocoa and *Pinus caribea* var. *hondurensis*. Longer internodes have been found under longer days in *Terminalia superba* and *Rauvolfia vomitoria*, among others, so one may conclude that either or both of the two components of shoot extension may be affected.

The effect of long-days therefore closely resembles that of warm nights (compare Tables 5.1 and 5.2), and it is interesting that in some cases the two factors interact together in promoting shoot elongation. Thus *Ceiba pentandra* seedlings produced 50 per cent more new leaves when *both* longer days and warmer nights were provided, but there was little effect when one or other treatment was given alone (Fig. 5.5). More complex interactions were involved with internode length and total shoot elongation, which were both strongly influenced by day-length and night temperature. (See also Plate 20A.)

The results of these experiments under controlled environments and those of Njoku (1964) show that a number of tropical trees are sensitive to photoperiod, and presumably therefore their rate of shoot elongation under natural conditions in the forest will be partly a function of this factor (see also section 3.1). Whether natural changes in day-length are sufficient to cause appreciable seasonal fluctuations in the rates of shoot growth remains to be seen. The variation at latitude 5° is a little over half an hour, while at 10° it exceeds one hour. The implication to be drawn from Table 5.2 is that growth rates can be increased by as much as 75 per cent/h; and it has even been claimed that some rice varieties are sensitive to changes of 15 or even 5 min! It may well be that Bünning (1947, 1948) was correct in his prediction that tropical plants would prove more sensitive to small changes in day-length than those of temperate latitudes.

The rate of shoot growth, influenced by a score of external and internal factors, is of obvious importance to the tree's ability to survive the competition in the tropical forest, and also to the primary production of forest or plantations. Equally important is the duration of growth, the period of time within which active extension and leaf production take place, and attention is now directed to the factors which may lead to the cessation of shoot growth.

5.3 Onset of bud dormancy

The ending of active outgrowth is signalled when no more new leaves are starting to expand at the shoot tip, and when stem extension is confined to a few internodes near the tips. Once this is completed, the dormant phase has been reached during which no further outgrowth or extension takes place (Plates 17B and 21B). Initiation of new leaves frequently occurs in the buds during the dormant period, but the

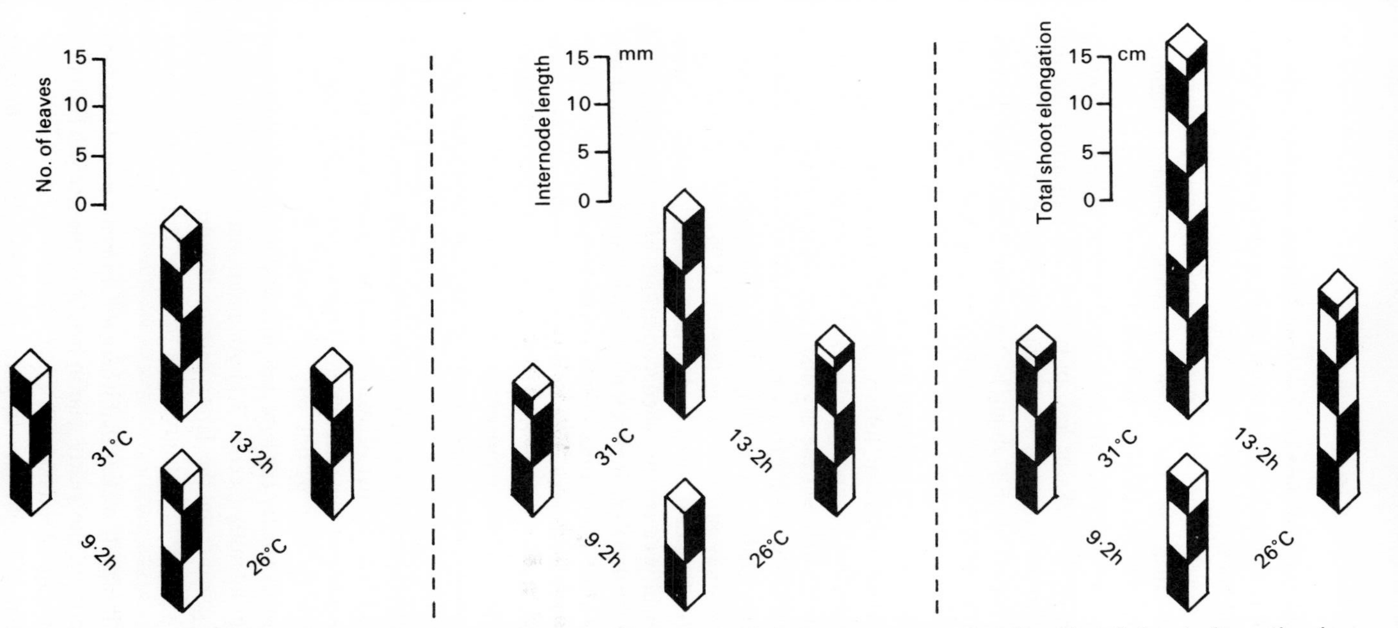

Fig. 5.5 Histograms showing interactions between day-length and night temperature upon the rate of shoot elongation in *Ceiba pentandra* seedlings. Note that the gain in height (right) may involve both greater production of leaves and longer internodes. The height of each column represents the gain during a 21-week period. Day temperature 31 ± 3°C.

initials remain small and clustered together. In some cases bud-scales are produced, while in others the terminal shoot apex is abscinded, leaving only the lateral buds.

Fewer ecological studies have been made of the time of cessation of shoot growth than on the more spectacular phase of flushing. Information on the duration of the period of active growth suggests that in cocoa, for example, it is about 6–7 weeks, while in rubber and mango it may be 2 weeks or less. In general, the indications are that trees other than seedlings and young saplings often make active shoot growth for a period of only about 1–2 months (Njoku, 1963; Alvim, 1964). In plants which flush intermittently or irregularly, therefore, the onset of dormancy will occur at different times of year, while in more regularly flushing species it will tend to be seasonal.

In the evergreen seasonal forests of West Africa the commonest month for the ending of shoot growth seems to be April, which is in the transitional period between the dry and rainy seasons (Table 5.3). It is rather remarkable that many tree species should be going dormant just as the 'growing season' is starting. Thus Greenwood and Posnette (1949) in their study of cocoa crops (see Fig. 5.2) observed that 'most of the growth of mature cocoa occurs in the drier months . . . while during the main wet season when conditions of water supply and humidity are most stable, little or no growth occurs'. A parallel may exist here with the many trees of the north temperate region which cease stem extension in May, June or July, when conditions for growth seem ideal. As Sachs pointed out as long ago as 1887, tree growth frequently ceases while temperature and other conditions favour vegetative activity, and resumes under conditions which are far less favourable.

Thus many tropical trees have a 'shoot growing season' which is shorter and earlier than that of many grasses and other crops. It is only possible to speculate on the selective advantages of such a life-cycle, but it may perhaps be that in this way newly-expanded leaves, at their maximum photosynthetic capacity, are produced in time for the period of greatest light intensity (section 3.1). An additional point may be that a leaf which expands in the latter part of the dry season may be subjected to less *overall* moisture stress than one formed at any other time.

Be that as it may, it is clear that where the dormant period is as long as 10 or 11 months, the yearly gain in height of the tree will be rather low even if the rate of extension is rapid. This partly accounts for the decrease in annual height growth with age – older trees of *Bombax buonopozense*, for example, are dormant for 9–10 months or more,

Table 5.3. Seasonal changes in the evergreen seasonal forests of Ghana. Generalised picture, based on semi-quantitative data. The horizontal lines indicate the time(s) of year when the event is most frequent; the dashes show moderate frequency. Vertical lines show the equinoxes.

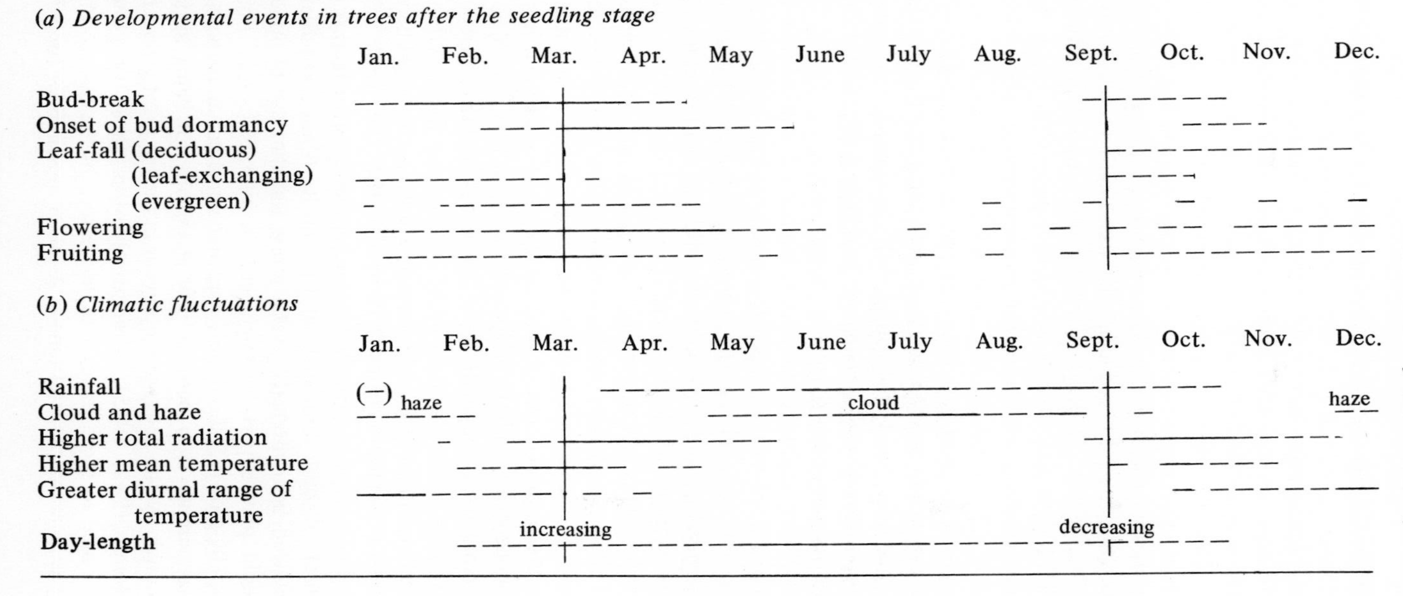

whereas 3 year old seedlings stop only for 3–4 months (Njoku, 1963, 1964) (see section 5.7). Similarly, seedlings of *Triplochiton scleroxylon* and *Hildegardia barteri* fell behind in height growth compared with *Musanga cecropioides* mainly because they stopped growing for a period while the *Musanga* grew continuously (compare also Fig. 5.3).

The physiological signals which initiate a change to the dormant phase appear to be both external and internal. Controlled environment studies have shown that shoot extension can be halted in a number of tropical trees by reducing the day-length (Njoku, 1964; Longman, 1966, 1969). Seedlings of *Hildegardia* stopped growth rapidly under 11½ h days and more slowly in 12 h days. With 12½ h, however, they grew continuously during the 26-week experimental period. Very rapid onset of dormancy also occurred in leafy *Cedrela odorata* plants exposed to short-days (Plate 21), but it took place much more slowly in *Terminalia superba* seedlings, about half the treated plants stopping temporarily, and the other half continuing to grow slowly. A proportion of *Ceiba pentandra* seedlings grown under short-days also showed temporary dormancy (Plate 19A).

Cooler night temperatures tended to induce bud dormancy in *Gmelina arborea*, some plants stopping with nights of 26°C, but none at 31° or 36°C (Longman, 1969). As predicted by Thimann (1962), temperature and day-length can interact, as in *Bombax buonopozense* for example. If these effects are general, it could be concluded that cooler nights and shorter days often lead to reduced shoot growth and to bud dormancy, while warmer nights and longer days tend to promote faster and more prolonged growth. Under natural conditions in the tropical forest, it might well be that cessation of shoot growth in some species could be influenced by these two factors, as is the case in a number of temperate trees. One could speculate that in West Africa naturally shortening days might perhaps stop the growth of *Hildegardia* seedlings in October, while temperature changes could be involved with larger trees stopping growth in April or May. (See Table 5.3).

However, it is probable that a number of other factors influence the onset of bud dormancy in the forest. Among external conditions, reduced light intensity, water stress and shortage of mineral nutrients could play a part, as might certain internal factors. For example, a rapidly-growing shoot may become depleted in carbohydrates, or it may simply stop because new leaves are not being initiated quickly enough by the apex, as may occur in the tea plant grown in the tropics (Bond, 1942, 1945). Alternatively, the 'flush' of growth may be confined to those leaves actually present in the bud, so that dormancy is as it were predetermined. Another possibility, for which there is experimental evidence, is that as the leaves on a growing shoot mature, they begin to inhibit further shoot extension. Successive removal of mature leaves in the intermittently-growing tree *Couroupita guianensis* delayed the formation of terminal buds so much that the plants grew continuously for 5½ months, in spite of the associated loss of carbohydrates and mineral nutrients (Klebs, 1926).

5.4 Growth of leaves

Young leaves often attract the attention in tropical forests because their colour is different from that of the mature foliage. Anthocyanin pigments in the cell vacuoles can make the leaves bright red or reddish green, as for example in *Cynometra ananta*, *Carapa procera* and cocoa. Other trees have very pale young leaves, while occasional cases have been reported where they are white or even blue (Richards, 1952). There has been a good deal of speculation as to the adaptive significance of these colours, but there is little clear indication what their function may be. The young leaves in some trees such as *Amherstia nobilis* hang downwards for a considerable time before assuming a more erect position, but again opinions differ as to whether this is due to lack of full turgor or undeveloped mechanical tissue.

An important feature is that in numerous species the orientation of mature leaves and leaflets can be altered by active leaf movements, brought about by growth curvature in the petioles, or controlled turgor changes in the pulvini. These are the swollen 'leaf-joints' which may occur at the leaf-base (see Plate 7B), and at various points in compound leaves (Funke, 1929, 1931). In many leguminous trees, the leaflets are folded together when emerging from the bud, but later they open in the day and close at night. In certain savanna species of *Bauhinia*, there are actually three separate movements in a leaf, caused by complex interactions between light and dark and internal timing mechanisms (Holdsworth, 1959). Similarly in the compound leaves of a large number of leguminous forest trees the leaflets can change their orientation in relation to the position of the sun, and also such factors as light intensity, time of day and moisture stress.

The physiological significance of changing leaf orientation probably lies in a balance between maximum photosynthesis and minimum water stress. As has been noted in section 4.2, it has important secondary effects in altering the leaf mosaic, and thus the proportion of light transmitted through the canopy at different times. The mosaic also changes markedly when a tree expands a large area of leaves in a short time, especially if it was leafless previously, but there are only slight modifications produced by leaf production in for instance a continuously-growing palm.

The time of year at which there are peaks of leaf expansion is the same as that for stem elongation, since the leaves commonly grow in size during a period of about 1–2 months following bud-break. An example from West Africa is given in Table 5.3. Stem and leaf growth are normally very closely linked; thus in the tea plant it has been noted that a single deciduous bud-scale amongst expanding foliage leaves has a corresponding short internode beneath it (Bond, 1942). On the other hand, young stems of *Dioscorea bulbifera* can grow very fast before leaf expansion starts, so this correlation is not always found. In most species of *Terminalia*, moreover, the first internode of a particular twig is very much longer than the later ones, although the leaf sizes are not very different. In fact, if the leaf surface on an expanding shoot of *T. ivorensis* is experimentally reduced, internode lengths are not significantly affected (Damptey, 1964).

Growth rates of leaves are often quite rapid, particularly during the middle part of the growth period. Leaflets of *Amherstia nobilis* can elongate at a rate of 18 mm/day and the petiole at 41 mm/day (Schimper and Faber, 1935). Leaf growth in this rapidly-flushing species may be completed in a fortnight or less, though it is another two weeks before the leaves spread out and stiffen (Funke, 1929). The rachis of the large fern *Angiopteris evecta* can even extend 90 mm in a day, and it is interesting that growth rates in these two species were lower in the day than in the night when the weather was sunny, but not on a cloudy day (Coster, 1927).

The final size of the mature leaf often depends on the age of the tree, and the length varies from a few millimetres to several metres. Leaf size is also influenced by various environmental factors, though not perhaps to the same extent as with the stem and root, since the leaf is an organ of determinate growth. A striking example of unequal growth of leaves is provided by those plants which show *anisophylly*; for example, the tropical American *Columnea sanguinea* (Goebel, 1928), *Anisophyllea trapezoidalis* of Malaya (Corner, 1940), and *A. laurina* of W. Africa. Here the branches have two rows of large leaves below and two of very small leaves above, which could perhaps be caused by the two types of leaf having very different growth potentials. However, since all the leaves on vertically-growing shoots are large, a 'gravimorphic' effect might be involved in inhibiting the upper-side leaves on the branches.

Water stress appears to reduce leaf growth of cocoa, for when the proportion of available water in the soil had fallen by a third, leaf growth had slowed considerably (Lemée, 1956). By the time the moisture supply had dropped by a further third, expansion had stopped entirely, with the result that the leaves only reached a length of 25–60 mm, while the control leaves were twice as long and still growing.

Shading also influences the growth of leaves of cocoa and coffee, with small, yellowish-green leaves produced in full sunlight, and increasingly larger and darker green leaves as shading is increased (Murray and Nichols, 1966). On the other hand, cocoa leaves in full sunlight grow considerably larger if the lower mature leaves on the same shoot have been removed (Krekule, 1972), so the effect of light intensity may be indirect. Even in rigorously controlled environments, in fact, it is very difficult to distinguish genuine control by light intensity from secondary effects of leaf temperature and water stress, while in field shading trials air and soil temperature, relative humidity and root competition from shade trees also have to be taken into account.

Other factors which influence the growth of leaves include mineral nutrition, on which it is difficult to generalise, although the effects can be very large; and temperature and day-length, which can interact together as they may do in some trees of higher latitudes (Nitsch, 1957). In *Triplochiton scleroxylon*, leaves on seedlings grown under 11 h days with nights at 20°C, were 35–40 per cent shorter than when the plants received either 14½ h days, or 30°C nights, or both (Plate 20B). *Ceiba pentandra* and *Gmelina arborea* also show interactions between day-length and night temperature.

5.5 Leaf senescence and abscission

The leaf typically reaches a peak of photosynthetic efficiency shortly before it stops expanding, gradually declining thereafter. Sooner or later a point is reached when it rather suddenly turns yellowish, or less commonly red, and is then actively shed from the tree by the completion of an abscission zone at its base. The change of colour is termed senescence, and is accompanied by breakdown of chlorophyll, ribonucleic acid and protein, and translocation out into the stem of organic and inorganic nutrients.

Very many factors can influence the leaf-fall of trees, including low light intensity, changed temperature and day-length, mineral deficiency, damage by mechanical means or pests, and water stress, though if the last is very pronounced the leaves may die without being abscinded.

Other considerations include the age of the leaf and the extent of competition with younger leaves or other growing parts (Addicott and Lynch, 1955; Kramer and Kozlowski, 1960).

Experimental studies on the leaf-fall of tropical trees are very scarce, and it is therefore not possible to draw any general conclusions. In *Bombax buonopozense*, ten times as many leaves were abscinded under short-days and hot nights as under long-days and cool nights (Table 5.4).

Table 5.4. Interaction of day-length and night temperature on leaf abscission in *Bombax buonopozense* seedlings. Mean number of leaves lost from each plant in a period of 4 months

	Short-days	*Long-days*
26°C nights	1·5	0·6
36°C nights	6·0	3·0

The above data suggest an interaction between temperature and day-length, but neither of these factors appeared to have any noticeable effect on several other species. There may, however, have been an indirect effect of day-length in *Hildegardia barteri*. More old leaves were lost under long-days in an experiment with this species, perhaps because they were competing with new leaves. Under short-days on the other hand, there was little competition since the plants were dormant for a long period.

An interesting case has been described in coffee plants, where a disease caused by the fungus *Omphalia flavida* leads to very early leaf-fall and greatly reduced yields. Here the abscission is probably not due to destruction of parts of the leaf by the fungus, but to disturbance of the hormone balance. Once the flow of auxin from leaf to stem is reduced, the abscission zone in the petiole completes its development, and the leaf falls off within a week (Sequeira and Steeves, 1954).

Leaf-fall occurs all the year round in tropical forests, but it is usual for more leaves to be shed at certain times. Studies of the seasonal production of litter, of which about two-thirds was leaf material, have shown that in ombrophilous forests in Colombia and Ghana, for instance, there was a maximum in March and a minimum in July (Bray and Gorham, 1964). In evergreen seasonal forest, there is frequently a peak of leaf-fall in the first half of the dry season as some of the trees become completely leafless (Table 5.3), while evergreen trees also may lose a proportion of their foliage at the same time (Beard, 1942). These seasonal fluctuations in leaf-fall have an effect both on the nutrient status of the soil and on the leaf mosaic and microclimate of the forest as a whole. In the life of the individual tree, the loss of leaves represents a reduction in photosynthetic and transpiring tissue, except when leaves are produced and lost steadily throughout the year.

How to distinguish an evergreen from a deciduous tree has been the subject of much argument and confusion. Basically, of course, it is a matter of the relative timing of bud-break and leaf abscission, but there are two other points to be considered. In the first place, the average life-span of tropical tree leaves varies from about 4 to 14 months, so that 'evergreen' means something rather different from the condition, for example, in north temperate evergreen forests, where leaves remain on coniferous trees for two years or more, sometimes even as long as seven years. Secondly, the shoot growth of tropical trees is not always synchronised, so that some branches may be in full leaf, others may be leafless, while some again may be in the flushing phase (Plate 18). Taking all these points into consideration, it may be useful to recognise four 'patterns of leafiness', although no very sharp lines can be drawn between such classes (see Fig. 5.6):

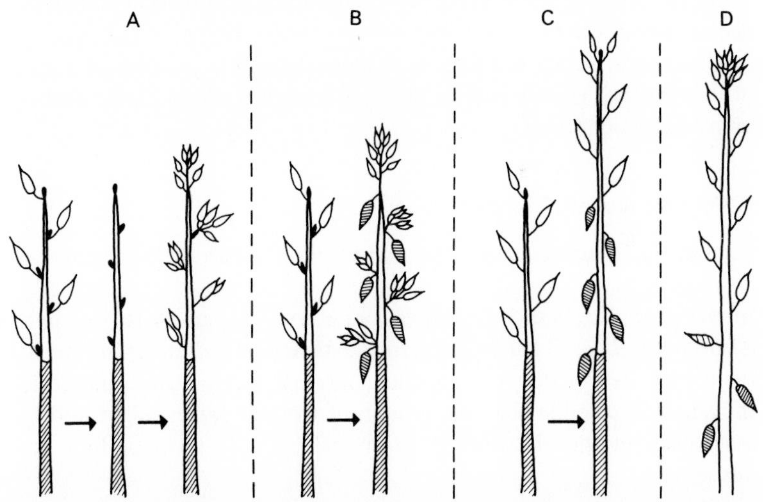

Fig. 5.6 Patterns of leafiness in tropical forest trees: A – periodic-growth, deciduous-type; B – periodic-growth, leaf-exchanging-type; C – periodic-growth, evergreen-type; D – continuous-growth, evergreen-type. (See description in text.)

A. Periodic growth, Deciduous-type:

Leaf-fall occurs well before bud-break; life-span of leaves approx. 4–11 months

In this case the branch or whole tree is leafless or nearly so for a period varying from a few weeks to several months. Individual trees can even remain completely leafless for periods exceeding 6 months (e.g. *Terminalia ivorensis*; Jones, 1967). Leaf-fall and bud-break are apparently not directly connected with each other.

B. Periodic growth, Leaf-exchanging-type:

Leaf-fall is associated with bud-break; life-span of leaves often approx. 12 or 6 months

Here flushing of new leaves starts at the time of the fall of the old ones, within about a week either way. A well-known example is *Terminalia catappa*, in which the leaves turn red, and then as they fall they are replaced by new (Plate 22B). In this species the process occurs twice a year, while in *Entandrophragma angolense*, *Dillenia indica*, *Ficus variegata* and *Parkia roxburghii*, for example, it happens once. It may well be that it is the senescence of the old leaves which provides the signal for bud-break (section 5.1), placing the leaf-exchanging type in a different category from evergreen or deciduous. This term should be distinguished from the expression 'leaf change' (German – *Laubwechsel*), which has been used indiscriminately to cover all stages of flushing, leaf growth and leaf-fall, and has given rise to a great deal of unnecessary confusion.

C. Periodic growth, Evergreen-type:

Leaf-fall is completed well after bud-break; life-span of leaves approx. 7–14 months

In this case, the branch or whole tree is truly evergreen, for example *Afrormosia elata*, *Celtis mildbraedii* and *Mangifera indica*. As in type A, there may be no direct connection between leaf-fall and bud-break, though it is possible that the growth of the new leaves might lead to senescence of the old, through hormonal control or competition for nutrients.

D. Continuous growth, Evergreen-type:

Continual production and loss of leaves; life-span variable, up to 14 months

Examples are *Trema guineensis*, *Dillenia suffruticosa* and the young seedlings of many forest trees. In these shoots there is, of course, no bud-break, but the rate of leaf production and abscission may each vary considerably, according to changes in the environment or competitive effects; and so the number of leaves present on the shoot may fluctuate.

The presence of substantial numbers of trees of type A in the tropical forest is a point of considerable interest. In evergreen seasonal forest in Ghana, for example, out of 158 species of forest trees about a third showed the deciduous habit (Taylor, 1960). There were some leafless trees to be found in every month except June, but with increased leaf-fall in the early part of the dry season, a peak in the number of species leafless occurred in December and January (Fig. 5.7). Flushing takes place in the majority of trees in February and March, and the number of species leafless falls to a very low value before the onset of the rainy season. (See also Table 5.3).

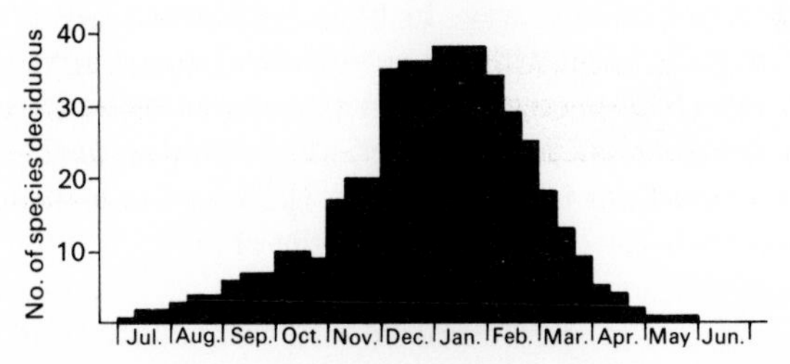

Fig. 5.7 Occurrence of the leafless phase in 158 forest tree species in Ghana. (Data derived from Taylor, 1960.)

There seems little doubt that the chief adaptive value of the deciduous habit lies in the avoidance of water stress in the middle part of the dry season. Yet it should not be forgotten that many other large trees of types B and C endure the dry season in a leafy state, just as many evergreen trees pass the cold winter months in forests of higher latitudes, while other species do so leafless. It seems that not one but several patterns have evolved in response to the problems posed by drought or cold.

In ombrophilous forests it is sometimes assumed that the deciduous nature of some species can be disregarded. However, as has been shown in section 3.3, the evaporative power of the air surrounding the crowns of the tallest trees may occasionally reach very high values, because of dry spells occasionally reaching areas normally moist and humid throughout the year. It is interesting to note that Richards (1952) states that in these forests the deciduous individuals occur almost exclusively in the upper layers, and that other trees, such as those with compound leaves, may shed a proportion of the leaflets at the same time, while the leaf as a whole does not fall off until later.

The frequency of occurrence of these four types of leafiness at different layers provides a basis for the physiognomic classification of the tropical forests (see section 4.5). The precise definition of such terms as 'evergreen', 'evergreen seasonal', 'monsoon' or 'semideciduous' varies from one system to another. Though the usefulness of the underlying character of leafiness is beyond doubt, it is necessary to be careful in comparison between different authors, and in the translation of terms.

5.6 Cambial and root growth

Another facet of shoot growth is the continued increase in thickness of the stem, and indeed most of the permanent tissue produced above ground is formed in this way. In time, individual tropical trees can become extremely large, with extensive crowns and root systems, and main stems which even exceed 2 m in diameter at the base. Some giant specimens are given in Richards (1952): notably a tree of *Entandrophragma cylindricum* in Nigeria which was over 5 m in diameter near the bottom. With total dry weights of the order of 100 tonnes, they rank amongst the largest living organisms.

Yearly increases in stem diameter vary enormously from shoot to shoot, and between different species. Middle-aged trees in the natural forest may add some 5–20 mm at the base of the main stem, while in very old emergent trees the amount is generally lower, sometimes less than 1 mm (Jones, 1956; Langdale-Brown *et al.*, 1964). Young saplings and pole-stage trees may also make slow radial growth, suppressed by the low light intensity and root competition of the larger trees, though they may be capable of rapid growth in gaps and along forest margins. Under plantation conditions also, younger trees can make very rapid diameter growth, average yearly increases of 20–60 mm having been recorded (Dawkins, 1967), with an exceptional figure of 90 mm for a two-year old crop of balsa (*Ochroma lagopus*) in British Honduras (Anonymous, 1960).

While there would be little problem in these latter cases, accurate measurement of the rate and duration of radial growth is generally rather difficult. Increase in thickness is the result of the activity of two lateral meristems, the vascular cambium and the phellogen or cork cambium, which are no doubt controlled in different ways. Moreover, successive girth or diameter measurements of trunks of trees do not necessarily show the total activity of the two cambia because, firstly, irregularities may appear in the bark, or whole pieces of it fall off. Secondly, the trunk may actually shrink appreciably during a hot day, due to the considerable water tension within it, so that it is really only early morning data which are comparable. The fact that xylem is under reduced pressure in the middle of the day can be elegantly demonstrated in certain lianes which have their crown in the upper tree layer. When a clean cut is made into the stem, one may hear a distinct sound as air is drawn in. This characteristic is also shown by 'simpoh' trees in the Dilleniaceae (Corner, 1940).

Measurements with an anatomical basis are therefore needed for accurate studies, and these are generally concerned with the major component, the production of xylem by the vascular cambium. In deciduous trees (type A; section 5.5), the cambium becomes dormant at the beginning of the leafless period, and starts to divide once more when flushing takes place (Coster, 1927–28). By careful measurements, experiments and anatomical investigations, Coster showed this connection between the leaves and the cambium to be a general one. The stimulus to renewed cambial activity in leafless plants must arise in the expanding buds, since the cambium remains dormant if they are cut off, or if a complete ring of bark is removed, isolating them from the portion of cambium under study. The influence of the expanding buds was not exerted through photosynthesis, for the cambium was still activated if the plants were kept in the dark, so Coster proposed a hormonal explanation at a time when plant growth substances had only recently been discovered.

It is very probable that it is the basipetal movement of auxin in the phloem from the expanding buds which is primarily responsible for the stimulation of cambial activity, though other hormones may also be involved (Morel, 1960; Wareing *et al.*, 1964). After leaf expansion is completed, it seems that sufficient auxin may still be produced by the mature leaves for cambial divisions to continue, but when the leaves senesce the supply dwindles and the cambium becomes dormant. Thus Coster was able also to cause the cessation of cambial activity by 'ringing' between the mature leaves and the part of the cambium being measured.

The timing of the period of cambial activity in a deciduous tree is thus controlled indirectly by the same external and internal factors which determine bud-break (section 5.1) and leaf-fall (section 5.5). One exception is provided by trees which flower during the leafless period, where opening flower buds and developing fruits can provide the stimulus for an earlier resumption of cambial activity. Nevertheless, there is still a definite period of cambial dormancy, and this is usually but not invariably marked by fairly sharp, continuous growth rings in the wood, though these are not necessarily annual. They may be due to changes in the size of cells or of tissues in the wood, or sometimes to the formation of a narrow tangential band of parenchyma or fibres as the cambium ceases or restarts activity (Coster, 1927–28; Lowe, 1968).

In the leaf-exchanging trees of type B, the position is not so clear-cut. In a tree of *Khaya grandifoliola*, for example, which was leafless only for a few days, the cambium did not become fully dormant, and only indistinct rings were formed (Hummel, 1946). However, in *Dillenia indica* and *Ficus variegata*, the cambium does become inactive, resuming when the new leaves expand (Simon, 1914).

Similarly in evergreen trees of type C, continuous cambial activity is probably the rule, but examples of dormancy have been claimed (Alvim, 1964). In the latter case, clear growth rings can be produced, but more usually they are indistinct, irregular in occurrence, or absent, especially in ombrophilous forest. The cambium can be rendered dormant experimentally by 'ringing' the shoot, so here too the mature leaves maintain a hormonal control. In evergreen trees of type D, including young trees showing continuous expansion of leaves, the cambium is almost certainly active throughout the year.

Whether cambial activity is continuous or periodic, there are often large fluctuations in the rate at which cells are formed and differentiated. Growth is often most rapid while shoot extension is proceeding, though there may be a delay at the base of a tree as a 'wave' of increased cambial activity proceeds down the branches and trunk. Sometimes the xylem formed while the leaves are expanding can be distinguished as 'earlywood' in contrast to the 'latewood' produced when the leaves are mature, but in other cases various arrangements of vessels, fibres, tracheids and parenchyma occur intermittently without any apparent seasonal pattern. In cocoa, cambial activity is low while flushing proceeds, and is at a peak during the rainy season (Alvim, 1964). Similar peaks have been recorded for other tropical trees by Hopkins (1970).

The factors which control the rate of cambial activity have been little studied experimentally, in spite of their obvious importance in forestry. Certain of these, such as temperature and water stress, can have a direct effect on the cambium, as does the force of gravity in branches and leaning stems. These and other external factors can also have an indirect influence on the cambium by affecting the leaves (Larson, 1964), while internal control systems such as the partitioning of assimilates between the various different meristems may play a part.

Very high temperatures, which reduced shoot extension in *Ceiba pentandra*, also reduced diameter increment significantly (Kwakwa, 1964), while a reduction due to water stress is suggested by a study of cocoa by Murray (1966). Trees given the equivalent of the annual rainfall (1,830 mm) spread evenly over the whole year showed about 30 per cent more diameter growth than those receiving natural precipitation.

Various types of tropical roots and root systems have been described in section 4.2, and conclusions drawn as to their ecological significance. Rather few root studies have in fact been made compared with those on the shoot system, due no doubt to the technical difficulties, but it is very important that more should be carried out, especially on an experimental, physiological basis.

Coster (1932) grew some 70 tropical, mainly woody, species at a site in Java which had a deep, permeable soil and regular rainfall. After 6 months, he dug out the seedling root systems carefully and found that in general the main root was longer than the main stem, and the spread of the horizontal surface roots was greater than that of the crown. Average root elongation rates were over 20 mm/day for the most rapidly growing trees, which is faster than that found in most temperate trees (Lyr and Hoffmann, 1967). The total length of the main root plus the primary or first-order lateral roots had actually exceeded 30 m in *Melia azedarach* and *Sesbania sesban*, though others had grown much more slowly.

Aerial roots have been used because of their convenience by several workers, and have provided valuable results, though these cannot necessarily be extrapolated to provide growth rates in soil. The daily course of root elongation was followed in aerial roots of a large liane, *Cissus adnata*, whose foliage was high in the forest canopy (Coster, 1927). The roots showed little or no growth when the sun was shining or when a drying wind blew during the night, but immediately after rain the rate of elongation rose temporarily to the very high value of 30 mm/h or more. Coster showed that the chief factor involved was the water status of the plant, and found that the roots could actually shrink by as much as 7 per cent in length when subjected to drying. In contrast, the aerial roots of *C. sicyoides* grew at a constant 4 mm/h throughout the night and a sunny day.

Seasonal changes in the rate of root growth no doubt occur widely, but have been little studied. Cocoa seedlings grown in glass-sided observation boxes showed peaks of root elongation several days prior to flushing, though in field studies the tap-root grew at a rather slow constant rate of about 15 mm/month over a period of 22 months (McKelvie, 1954; 1958). There may also be a correlation between flushing times and the rate of root growth in the tea plant (Wight and Barua, 1955), and it may well be that root and shoot growth are linked by internal control systems as well as being influenced by their separate external environments.

Another gap in our knowledge of root growth is whether it ceases entirely at certain times of year. As far as is known, tree roots do not possess any inherent dormancy in the sense that buds and seeds may do, and they appear to stop growing in temperate regions because of unfavourable temperature or moisture conditions, or because they are not receiving a stimulus from buds and/or leaves (Lyr and Hoffmann, 1967; Richardson, 1957). Cambial activity in roots of tropical trees has only been investigated in relation to the formation of buttresses.

The initiation of new roots takes place not only within the existing root system but sometimes also by the formation of root primordia in stem tissue. These may grow out as adventitious roots (see Plates 10 and 13A), forming aerial root systems of various kinds, and sometimes penetrating into soil or into crown humus (section 3.4). A considerable number of tropical forest species are capable of forming new root systems from detached shoots, or even sometimes leaves, but it is an open question as to how far this occurs under natural conditions; and the same applies to the capacity of root systems of *Chlorophora excelsa* to regenerate buds.

5.7 Physiological changes with age

It has already been remarked that young seedlings and older trees often show different growth forms. A striking example is the presence of thorns on young seedlings and coppice shoots of citrus trees, and their absence from older, flowering stems, but the modifications with age are generally more gradual, involving, for example, the size, shape and arrangement of leaves, or the length of the growing season. These quantitative changes can result in leaves taken from the top of an emergent tree being very unlike those of a seedling of the same species growing on the forest floor, and they are frequently a good deal smaller.

It is one thing to recognise that such variation occurs, but quite another to determine its cause, for the leaves from the emergent tree could be smaller for one or more of three contrasting reasons. Firstly, the external environment is very different in the two parts of the forest (Chapter 3), so that such factors as light intensity, temperature or moisture stress could be responsible. While it is very difficult to test this experimentally for technical reasons, it is interesting to note that seedling leaves may still be larger than those of emergents, even when the young tree is growing in a large gap or clearing, where the microclimate is much closer to that at the top of the canopy.

Secondly, the leaves are borne on trees of very different sizes, with many more competing sites for growth and storage in the big tree, and with greater distances for translocation. This problem has been approached experimentally using the simpler shoot system of cassava, a short-lived woody plant with regular and repeated increase in the number of growing points, and associated decline in vigour of growth (see Fig. 4.6C, p. 60). Removal of all growing points but one led to a stimulation of its rate of growth, and leaves were 35 per cent longer than in unpruned controls, so it is evident that competition can indeed influence leaf size (Damptey, 1964).

In the third place, there may be an effect brought about by the physiological age of the trees. It is possible to eliminate size and microclimate differences by striking cuttings or grafting small shoots from big trees on to seedling root-stocks. When such vegetatively-propagated plants are grown in the same environment as potted seedlings of similar size, they frequently exhibit a different growth habit (Plate 24), and may flower sooner and more heavily. It seems that during the life of the old tree certain changes have occurred that are retained rather firmly by the tissue, which may then be termed 'mature' in contrast to the 'juvenile' condition found in seedlings, and also in coppice shoots arising near the base of the trunk of an older tree.

Differences clearly due to physiological age were found in similar-sized mature grafts and juvenile seedlings of *Cedrela odorata*, grown in growth rooms under different photoperiods (Longman, 1969). The rate of shoot elongation was more strongly influenced by day-length in the mature plants, which also ceased extension for a period under both long- and short-days, whereas the juvenile trees only became dormant under short-days (see Plate 21). On the other hand, Plate 25A illustrates differences which are definitely not due to calendar age, since all the trees are 5 years old. For reasons of external environment and/or size, the tree which grew rapidly has entered a phase where it has distinctly periodic shoot growth, with a deciduous period followed by bud-break; whereas the slower-growing trees were evergreen and grew continuously for long periods. Whether the big tree has become physiologically older remains an intriguing but open question.

It is generally impossible to determine which factors are at work in the uncontrolled setting of the tropical forest, but a number of morphological differences between the foliage of large and small trees have been widely reported (Taylor, 1960). Big trees very often have smaller leaves, as has already been noted, and compound leaves may have fewer leaflets, notably in the Meliaceae. As trees become larger, they often stop producing leaves continuously, internodes may shorten, dormant periods lengthen, and the leaf-exchanging or deciduous habit may appear. In addition, there is a tendency for individual plants of the same species to become more synchronised in their development when they are larger, as can be seen for instance by comparing the out-of-phase condition of a 3- or 4-year old plantation of *Terminalia ivorensis* (Plate 23A) with the comparative regularity of one about 20 years of age.

As a seedling tree becomes larger, the characteristic branching patterns also emerge (see Fig. 4.6). In *T. ivorensis*, for example, there is a rapid but short-lived extension of a tall leading shoot bearing large leaves (Plate 23B), followed by a more prolonged branch growth period. This consists of brief bursts of elongation, repeated over and over again by increasing numbers of sympodially growing branchlets, giving the characteristic 'pagoda' shape found in this genus (Corner, 1940). Experiments have indicated that in *T. ivorensis* seedlings there are effects of laterals upon the leading shoot as well as vice versa (Damptey, 1964), so that the dominance relationships of the whole shoot system must certainly be complex. Among the many external factors which may influence branching, day-length has been shown to affect *Rauvolfia vomitoria* seedlings (Piringer *et al.*, 1958), while cocoa responds to temperature as well (Piringer and Downs, 1960; Murray, 1964). On the other hand, Hallé and Martin (1968) have reported that a clone of rubber trees usually formed the first branches after the terminal shoot had made nine flushes of growth, which suggests a strong measure of internal control.

Amongst the many other features of larger trees may be mentioned a variety of changes in bark characteristics, and the formation of buttresses in some species by an increase in the number of cambial initials in localised areas of the stem and root, which produce quantities of xylem rather different from the normal (Stahel, 1971).

Probably the most fascinating of all the changes in the life of a tree is the development of the capacity to flower. The initiation and development of floral structures will be considered in the following section, but we are concerned here with the onset of the reproductive phase, which often does not occur until the tree is quite large. Occasionally, precocious individuals may produce a few flowers before they are 2 years old (e.g. *Trema guineensis*, *Hildegardia barteri*, *Funtumia africana*, *Monodora tenuifolia*). Cocoa may also start flowering early in life, though not until main stem growth has stopped and sub-terminal 'fan-branches' of a different growth habit have been produced.

In most trees, however, the 'juvenile period' through which they have to pass before they flower abundantly is longer. For example, more than half the trees in some plots of a teak provenance trial near Ibadan, Nigeria, had commenced flowering 5½ years from planting, though it was noticeable that trees originating from certain localities were still mainly vegetative. In this important timber tree, the large inflorescences are terminal on the main stem as well as on branches, and the onset of the reproductive phase determines in part the yield of straight logs that can be obtained. Records of dipterocarps in the arboretum at Kepong, Malay peninsula, suggest that the age of first flowering is usually 20–30 years, and sometimes more (Ng, 1966). An unusual case is that of the monocarpic plants such as certain bamboos which flower only after a large number of years; they may do this gregariously (section 5.8) and then all die. What may be the factors which precipitate such behaviour remain largely conjecture, for the evidence is very sparse.

Short-days hastened the onset of flowering of 18-month old coffee seedlings (Piringer and Borthwick, 1955), while removing a complete ring of bark from citrus shoots advanced the time of first flowering from 5–10 years to 3 years (Furr *et al.*, 1947). It is not known why ringing has this effect, which is also found with a number of temperate forest trees. Some indication of a possible hormonal effect is suggested by the stimulation of flowering in 10-year old cocoa trees, that had never been seen to flower previously, by transplanting into their bark small flowering cushions from very floriferous trees (Naundorf, 1954).

The onset of the reproductive phase is clearly of great importance in the regeneration of the different forest species (compare section 4.3). There seems little doubt that environment, size and age will all play a part, but much more study is required to determine this. Perhaps the best studied species in the tropical forest is the herbaceous *Geophila renaris*, which remains vegetative as long as the soil moisture content is near its maximum value. If it falls below the permanent wilting percentage (see Fig. 3.7, p. 35) flowering is stimulated, and it continues afterwards even if the soil again becomes saturated with water (Bronchart, 1963). Such an experimental approach could show whether or not water stress also induces any forest trees to start flowering.

5.8 Flowering

Flowers in the tropical forest have often attracted attention because of the profusion of different shapes, sizes and colours which occur. *Rafflesia arnoldi* has flowers more than a metre across, which are found on the forest floor; in cocoa the hundreds of small flowers are borne on the trunk and larger branches; *Bombax buonopozense* produces a red, waxy 'pond', in which various aquatic organisms may flourish, high up in the upper canopy. It is obviously very difficult to generalise about cases as different as these, and it is also harder to experiment on flowering than on vegetative growth; but certain general principles apply.

The first stage of reproduction is the initiation of flowers or inflorescences by certain shoot apices, often those towards the top of the tree. Continued development and opening of the flowers can occur without a break in a few woody species (e.g. *Hibiscus*), just as is usually found in herbaceous plants, but in most trees there is an interval, sometimes as long as 9–12 months, during which the flower initials remain within a dormant bud. Care is therefore needed to distinguish between the timing of and conditions promoting, floral initiation, and those concerning the opening of the flowers.

Continuous flowering all the year round occurs in some shrubs and a few continuously-growing, evergreen trees, such as *Trema guineensis* in Ghana and *Dillenia suffruticosa* in Malaya. Trees of the latter may even flower uninterruptedly for about 40 years (Corner, 1940). However, the great majority of species have a definite flowering season or reproduce irregularly, but the tropical forest is never without flowers, because of the large number of different species and variation between individuals in the time of flowering. In some cases, indeed, flowering can be seen in parts of the crown of a particular tree, while other branches are in a different phase (Plate 18). On the other hand, it is also quite common to find synchronised or gregarious flowering, as for example in *Pterocarpus indicus* in Malaya (Holttum, 1953), *Tabebuia serratifolia* in Surinam (Schulz, 1960), and the West African liane *Calycobolus (Prevostea) heudelotii* (Rees, 1964), where every specimen of the species is in flower at the same time, sometimes over many hectares of forest.

The interval between successive flowering periods is variable, some plants being non-seasonal, while others flower regularly once, twice or even more frequently in a year (Koriba, 1958). A few of the annual type are so regular that their flowering has been used as a signal for the planting of a crop plant; for example *Erythrina orientalis* for yams in the New Hebrides (Baker and Baker, 1936), *Sandoricum koetjape* for rice in Malaya (Corner, 1940), *Trichilia heudelotii* for the second planting of corn in Ghana (Irvine, 1961).

Flowering can also be biennial, as in some varieties of mango, or less frequent still, in which case the scarcity of flowering years can raise practical problems of seed supply, as for example for forestry plantations of *Triplochiton scleroxylon*. On the other hand, such infrequent flowering is apparently sufficient to provide for regeneration of the species in the natural forest, even in extreme examples like *Homalium grandiflorum*, which flowers only once in 10–15 years in Singapore (Holttum, 1953), or in the monocarpic plants mentioned in section 5.7.

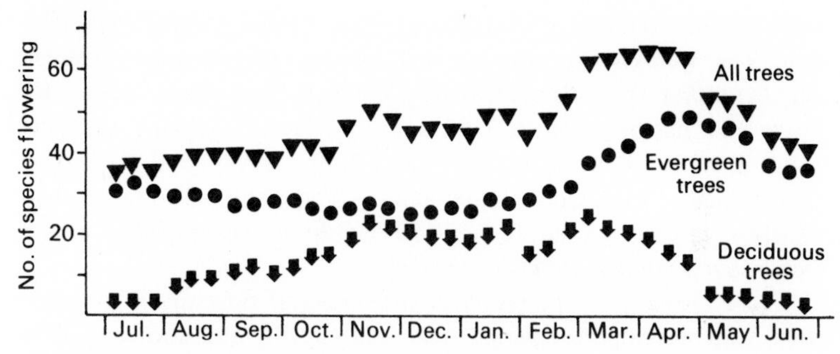

Fig. 5.8 Frequency of flowering by months in 158 deciduous and evergreen tropical trees in Ghana. (Data derived from Taylor, 1960.)

Considering the forest as a whole, there are usually peaks of flowering at certain times, even in near-uniform climates (Richards, 1952). In evergreen trees, these peaks are most commonly found during the transition from a drier to a wetter season (Fig. 5.8), with terminal or lateral inflorescences or single flowers being borne on the new shoots which are often produced at this time (see Table 5.3). In deciduous trees in evergreen seasonal forests, there may be a tendency for flowering to be more common for a period during the dry season, the flowers either emerging from separate flower buds which do not contain foliage leaves (e.g. *Ceiba, Hildegardia*), or alternatively emerging together with the newly-flushing leaves (e.g. *Delonix, Peltophorum*).

The factors leading to emergence of the flowers have been little studied, but in the latter case they are likely to be the same as those which cause bud-break in purely vegetative buds (see section 5.1). Separate dormancy relationships presumably apply to flower buds which open at a different time from the corresponding leafy buds. A number of experimental studies have been made on the coffee plant, in which the flower buds generally remain dormant when soil moisture is made freely available, for instance by irrigation. The dormancy can be broken by a period of 3–5 weeks of dry conditions leading to leaf moisture deficits of at least 10 per cent (Boyer, 1969), and all the crop will flower simultaneously about 10 days after the water stress has been relieved (Alvim, 1960). An alternative method of breaking the dormancy is by a sudden reduction in temperature (Coster, 1926; Mes, 1956–57), which has to be at least 3°C in 45 min or less. Evidence has been obtained by Browning (1971) which suggests that changes in levels of growth promoters and inhibitors may account for these effects.

It often happens, of course, that there is a small but rather sudden drop in temperature when there is a heavy shower, so that either or both factors could be responsible in other species which flower after rain. The well-known case of 'rain flowers' in the Amaryllidaceae (*Zephyranthes* spp., *Pancratium* spp.), which flower three days after heavy rain, has been shown to be a direct effect of moisture, since the addition of either warm or cool water after a dry period is effective (Holdsworth, 1961). On the other hand, a temperature-drop rather than moisture appears to be the critical factor in the epiphytic orchid *Dendrobium crumenatum* (Coster, 1926).

The ecological interpretation of the timing of flowering is complex, since fruit development periods and seed viability and dormancy have to be considered separately for each case. One important consideration is the effective display of the flowers such that they attract insects and other pollinators, and this could be related in some cases to the occurrence of flowering when the branches are leafless or with leaves only partly grown. The synchronisation between pollen shed, receptivity of the stigma and the period of activity of the appropriate adult insect is often very close (see Faegri and van der Pijl, 1966). For example, flowers of *Ceiba pentandra* open as dusk falls, are pollinated by flocks of bats which lick the nectar and distribute pollen on their fur (Baker and Harris, 1959). By the following afternoon, the corollas are falling, so that the flower does not last a full day. Many other tropical tree flowers are similarly short-lived.

The timing of floral initiation has hardly been studied, although it is fairly easy to determine by dissection if specimens can be obtained, for instance, at fortnightly intervals. In *Monodora tenuifolia* and *Bosqueia angolensis*, Njoku (1963) noted that the buds which will contain floral as well as leaf primordia can be detected from their larger size in October, 4 or 5 months before flowering occurs. In *Bougainvillea glabra*, a widely-cultivated ornamental shrub, the process can be followed rather simply, since an accessory lateral branch becomes a thorn in the vegetative condition and a peduncle when flowering occurs (Hackett and Sachs, 1966).

The physiological factors which cause growing points of tropical trees to become reproductive have been little studied. Some plants of economic or ornamental value have been investigated, and it is known, for instance, that coffee, tea, bougainvillea and poinsettia are all short-day plants. The critical day-length below which flower initiation occurs has been estimated at 11¼ h in tea (Barua, 1969), and 12 h or less in coffee, though a few flowers may be initiated in continuous illumination, so the inhibition is not an absolute one (Went, 1957). In *Bougainvillea glabra* also, some flower initiation takes place under longer day-lengths, though this depends also upon the temperature (Hackett and Sachs, 1966). Similarly in the decorative shrub *Euphorbia pulcherrima*, flower initiation could take place at a day-length in excess of the usual critical value if the plants were kept at a constant 15° or 22°C, or if 12 out of the 24 hours was spent at 15°C and the rest at a warmer temperature (Kwakwa, 1964).

The stimulation of flower initiation by 'ringing' in mango (Mallik, 1951), cassava (Damptey, 1964) and in *Faurea speciosa* (Noel, 1970), indicates that internal conditions are also important. Little is known about these, however, nor about the interaction of internal condition with external environment, which presumably determine floral initiation, and thus indirectly flowering, fruit and seed production, and the regeneration of the tropical forest.

5.9 Fruits and seeds

With the exception of parthenocarpic plants, such as the banana, pineapple and seedless varieties of citrus, pollination must usually occur for fruit development to proceed. Thus if cocoa flowers are unpollinated, they shrivel and fall off within three days, while in pollinated flowers the ovary continues to expand steadily, and the fruit is set (McKelvie, 1956). However, fruit setting evidently involves other factors as well, since there are up to 6 000 flowers on a single inflorescence of mango, for example, yet only 2–4 of them normally set fruit (Singh, 1960). It is not clear what are the factors concerned, but it has been suggested that nutritive and hormonal systems may be responsible, and that flower abscission can also be caused by water stress. Fruit-set in coffee was reduced or absent at high and low temperatures and greatest at 23°C day/17°C night and at 26°C/20°C respectively (Went, 1957).

Shortage of soil moisture has been shown also to reduce the rate of enlargement and final size of many fleshy fruits, and there are even measurable reductions in fruit size at mid-day due to rapid transpiration (Zahner, 1968). Temperature affects the rate of fruit growth in coffee, which was much quicker with a 26°C day/20°C night than a 17°C/12°C respectively (Went, 1957).

Not every fruit survives to maturity, some being broken off or eaten by various animals while many others are actively abscinded, perhaps because of competition with other developing fruits or expanding leaves. Cocoa fruits are liable to a condition known as cherelle wilt, in which 70–90 per cent or more shrivel and remain on the tree. It is not clear what causes this phenomenon, which is commonest around 7 weeks after pollination and again at 10 weeks, though it is increased by complete 'ringing' of the fruit stalk (McKelvie, 1956), and auxins appear to be involved (Krekule, 1969).

The final stages in fruit enlargement are usually rather slow, but then the ripening process often occurs quite rapidly. In many fleshy fruits, there is a sudden rise in the rate of respiration within the fruit, which is associated with the metabolic changes leading to softening and the development of the characteristic colour and taste. The onset of ripening appears to be controlled partly by ethylene levels in the fruit, and it can be hastened by addition of ethylene or by picking the green fruit, as in avocado pear and pawpaw.

Most of these fleshy fruits as soon abscinded, or break off and fall to the ground, where they may be dispersed by various animals, while others may be eaten *in situ* by birds, bats, monkeys, etc., and the seeds scattered over a wider area. As noted in section 4.2, wind dispersal occurs in a number of emergent tree species, but in few of the species of the lower storeys. In many non-fleshy fruits, the ripening stage involves drying and splitting open, so that the seeds are dispersed while the fruit remains on the tree.

There are fruits growing and ripening in the tropical forest all the year round, though in Ghana, for instance, there are many more trees to be seen fruiting during the dry than in the rainy season (Fig. 5.9). Fruiting is most common in March, and least frequent from May to July. Some of these fruits develop from flowering which occurred two or three months previously in the same dry season (see Fig. 5.8), but some represent slower-growing fruits from flowering up to about 10 months previously.

Individual tree species may have a short fruiting season of one or two months only, as for example *Lovoa klaineana*, or fruiting may be spread over as much as 8 months of the year (e.g. *Mimusops heckelii*). The continuously flowering species, such as *Trema guineensis*, produce fruit all the time, while species flowering twice a year may also fruit twice. The latter group include *Bosqueia angolensis* and *Celtis* spp. in Ghana (Taylor, 1960), and coffee plantations in Kenya, where there is usually an early crop from January to July and a main crop from June to November (Huxley, 1970). Even though a species may flower each

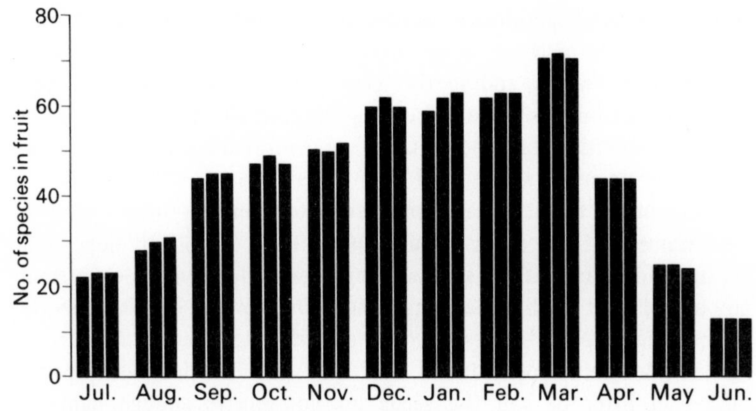

Fig. 5.9 Fluctuations in fruiting through the year in 158 forest trees in Ghana. (Data derived from Taylor, 1960.)

year it may fruit less frequently, and it is reported that biennial fruiting is characteristic of all trees of Vochysiaceae, and the majority of Lecythidaceae, Burseraceae and Leguminosae in a rain forest in Surinam (Schulz, 1960). In general it seems that fruiting tends to be rather more variable in its timing than the other seasonal changes in the forest (see Table 5.3, p. 95), presumably because of the influence of a variety of factors on the progress of fruit development. Thus fertile *Terminalia ivorensis* fruits have been collected in most months of the year in various parts of West Africa (Lamb and Ntima, 1971), but on the other hand the fruiting of *Claoxylon hexandrum* is so regular that it is used to mark the date of a festival (Irvine, 1961).

The size of tropical tree seeds varies from those which weigh less than a milligram (e.g. *Musanga cecropioides*), to others such as *Mimusops heckelii* weighing 20–30 g. Once they have reached the soil surface, they may germinate immediately, as in cocoa, many dipterocarps, *Mora excelsa* and *Montezuma speciosissima*. In many species, such as *Khaya ivorensis* and *Brachystegia laurentii*, it has not been found possible to dry and store the seeds. In *Aucoumea klaineana*, viability dropped from 80 per cent on collection to 15 per cent after two weeks' storage (Becking, 1960). On the other hand, cold storage allowed *Swietenia macrophylla* and *Cedrela odorata* to be kept for 6–12 instead of 1–2 months (Marrero, 1943), and *Terminalia ivorensis* for about a year with little loss of viability (Olatoye, 1968).

Under forest conditions, hard-shelled seeds in indehiscent pods (e.g. *Albizzia lebbek* and *Hymenaea courbaril*) may remain in good condition for several months (Marrero, 1942). Such 'hard' seeds, which are quite common in the Leguminosae, are frequently dormant until mechanical abrasion or the activity of micro-organisms has rendered the coats permeable to water, and in this case it is easy to produce a high percentage germination just by rubbing the seed on a file, or in extreme cases by immersing for a short time in concentrated acid. Other cases of dormancy are more subtle, and may explain some of the conflicting evidence in the literature, and irregular performance in forest nurseries.

A number of tree seeds are now known to be completely dependent on light for germination, so that they remain dormant unless a small amount of light reaches the living tissues. *Chlorophora excelsa*, *C. regia*, *Nauclea diderrichii*, *Mitragyna ciliata* and *Musanga cecropioides* are all light-requiring (Olatoye, 1970), and it is possible that when partly buried in litter or soil the seeds do not receive enough light for germination unless the canopy has been broken. There are in fact dormant seeds of various secondary forest species present in the surface soil of primary forest which has probably never been cleared or disturbed by man (Symington, 1933), and a light requirement may be one of the factors involved in their ecology.

In other cases, seeds may require 'after-ripening' for a period before full germination is possible. In rice and many other cereals the dormancy of freshly-harvested grains can be removed by a few weeks' dry storage at 40°C, and the same is true of oil-palm (Rees, 1961). On the other hand, storage at 50 and 70 per cent relative humidity at 22°C led to the development of a secondary type of dormancy in *Hildegardia barteri*, which was not the case when the fruits were stored at 90 per cent relative humidity (Enti, 1968). This effect could be largely removed by placing the drier seeds into a moister atmosphere, and it apparently involves changes in the permeability of the structures surrounding the seed itself. It may perhaps be that this mechanism operates under natural conditions on the steep, rocky slopes colonised by this species (see Plate 25B). During the dry season, the majority of seeds may perhaps remain unresponsive to occasional rain or dew, but in the moister conditions of the rainy season they may lose their dormancy, and germinate with a greater chance of survival.

The temperature at which germination itself takes place is also important. Generally speaking, the quickest germination occurs at quite high temperatures: thus germination in coffee took three months at 17°C, but only three weeks at 30°C (Went, 1957). A temperature of 30°C was also around the optimum for the five species studied by Olatoye (1970), although some tropical seeds can be inhibited from germinating at temperatures as high as this, while being promoted by diurnal fluctuations in temperature (see Longman, 1969).

Germination is thus a complex process, which often takes place at widely different times even in seeds derived from a single fruit. Seed dormancy appears to be an important survival mechanism in the tropical forest, leading to specific but diverse timing of germination. Many seedlings still fail to survive, of course, but the chances are increased of a few doing so, when the conditions happen to be more favourable.

Chapter 6

The future of the tropical forest

Reference has been made several times in these pages to the great bulk, luxuriance and potential of the tropical forest. Comparison of the total organic matter or biomass with other vegetation types (Table 6.1) shows that the dry weight of plant material per unit area in tropical forests is generally very high. On a world basis its biomass is clearly the greatest, actually representing about a half of all living matter.

Great emphasis is placed nowadays on the net primary production which a vegetation type or crop can produce. It has been estimated that tropical forests are often producing between 10 and 50 tonnes of dry matter per hectare every year, and this again is generally higher than other vegetation types, except perhaps a few particularly high-yielding agricultural crops and forestry plantations. The claim that tropical forests represent a unique world natural resource seems to be fully justified, for they produce some 4×10^{10} tonnes of dry organic matter every year, more than any other source.

From the point of view of its usefulness to man, this vast stock and annual production is of uneven value. The inhabitants of the forested regions of the tropics use the diversity of tropical forest plants to its full. Building materials and tools, roof coverings and wrapping materials, cloth and ropes, medicines, food and fodder for the animals all come from particular species collected or grown for the purpose. No wonder, therefore, that the life of the forests is known intimately, and passed to succeeding generations by oral tradition and legend, so that ecologists from overseas can discover among villagers tree-spotters who can identify hundreds of species.

Nevertheless, not more than a tiny fraction of the annual production of the forest is utilised to meet these needs, and moreover some of the forested land is usually felled, burnt and farmed to provide a carbohydrate supply (see Plate 26). As populations increase, it is easy to see that the extension of farming continually reduces the supply of the forest products. In addition, as we have seen, with the removal of biomass go far-reaching changes in the soil, and loss of the very environment which made the high productivity possible. As a result, agricultural yields, which are often rather low, particularly in Africa (see Phillips, 1959), quickly fall and deforestation proceeds faster still.

Table 6.1. Net primary production and biomass estimated for different types of vegetation

		Net primary production		*Biomass*	
	Area (10^8 ha)	*Per unit area* (dry tonnes/ha/yr)	*On a world basis* (10^9 dry tonnes/yr)	*Per unit area* (dry tonnes/ha)	*On a world basis* (10^9 dry tonnes)
Tropical forest	20	20(10–50)	40·0	450(60–800)	900
Savanna	15	7(2–20)	10·5	40(2–150)	60
Temperate forest	18	13(6–30)	23·4	300(60–2 000)	540
Boreal coniferous forest	12	8(4–20)	9·6	200(60–400)	140
Tundra and alpine grassland	8	1·4(0·1–4)	1·1	6(1–30)	5
Steppe and other temperate grassland	9	5(1·5–15)	4·5	15(2–50)	14
Agricultural land	14	6·5(1–40)	9·1	10(4–120)	13

Data from Whittaker (1970)

It is obvious that the population must usually obtain the majority of its food locally, and it is therefore of the greatest importance to plan a reasonable pattern of land-use to take account of present and future requirements. In each district, the areas most suitable for cultivation and plantations need to be determined, with other portions reserved under forest in order to provide on a sustained basis food, materials for local industries and export, and other forest produce.

In this connection the alluvial soils in the lower part of catenas and on flood-plains are potentially more suitable for permanent agriculture, being richer in nutrients and less prone to loss of fertility or erosion when cleared. However, except for rice cultivation, there are considerable problems of drainage and health which have to be overcome. In parts of the tropics, irrigation schemes and drainage systems have allowed a considerable extension of the area and range of crops. Various combinations of agriculture with forestry appear to be promising, but newer methods have to be tested carefully by pilot trials, to avoid the failures which frequently attend more grandiose projects (see Gourou, 1953; Dumont, 1966). Tree crops appear to be more suited to the humid tropics, particularly on mesic sites with ferralitic soils. Considerable success has been achieved locally with such crops as oil-palm, rubber and citrus, and with cocoa and coffee. In none of these, however, is more than a small proportion of the primary production used, and being cash crops for export they are subject to externally controlled fluctuations of price. A more stable future may lie with timber growing as a basis for the development of local industries (that is, an intermediate technology), and also for foreign sale. Dawkins (1964, 1967) has drawn attention to the very high yields which are possible with certain species in the tropics and subtropics. Production values between 5 and 15 tonnes dry weight of above-ground wood per hectare per year have been estimated for plantations of *Aucoumea*, *Dryobalanops*, *Terminalia*, *Shorea*, *Tectona*, *Ochroma*, *Cedrela* and *Triplochiton*; while *Swietenia* spp. can sometimes produce up to 20 tonnes/ha/yr.

Such rapidly growing plantations are presumably very efficient in utilising the sun's radiation, which broadly delimits the productive potential of an area. It may well be that certain species have a much greater photosynthetic capacity than others. However, Coombe and Hadfield (1962) concluded that the rapid growth of *Musanga cecropioides*, which has been mentioned several times in previous pages, does not lie in a particularly efficient net assimilation rate. The most important consideration appears to be the unrestricted production of new leaves, and therefore the duration of the shoot elongation period (see section 5.3) is likely to be a critical factor in primary production.

Ceiba pentandra, on the other hand, appears to have a strikingly high net assimilation rate for a woody plant (Okali, 1971).

In addition to providing timber for local use and export, a number of fast-growing trees are suitable for making paper pulp, and for other newer uses of wood. It should not be overlooked, moreover, that advances in biochemistry and technology could change the patterns of wood utilisation, perhaps allowing new species to be added to the commercially important materials, or possibly moving wood into the position of a major energy-rich substrate for industrial processes. Yields in present-day terms may also change considerably as genetical selection and tree improvement proceed. For example, the international provenance trial of *Cedrela* is already indicating that very large differences in early growth rate can occur between trees originating from different geographical areas (see Plate 17B).

As well as mixed agricultural and plantation land-use, it is in the interest both of local and world economies that some forest should be retained. It should not be imagined that this would always be a kind of nature reserve, kept for scientific interest alone. Methods are known, for instance, of group and strip enrichment planting, which can increase the proportion of specially valuable timber trees; while more sophisticated silvicultural approaches are discussed by Lamprecht (1961). To a reasonable extent, fodder for stock animals, leaves for protein extraction, or bush-meat can be removed without undue disturbance of the forest ecosystem. Forest reserves are also particularly appropriate in steep terrain to prevent erosion and rapid run-off of water in catchment areas, and these have been recognised for many years as 'protection forests'.

There are many ecological reasons for retaining a sizable proportion of the total area of tropical forests. It would be tragic for both science and the economies of the tropical countries if such a large and varied resource was simply allowed to dwindle away, perhaps being replaced by farm-bush or savanna, consisting mainly of woody weeds and grasses. Apart from any other considerations, species of plants and animals might become extinct before the world at large had recognised their existence, nature or usefulness.

Most damaging of all, perhaps, would be the loss of these diverse communites with all their range of structure and function. The tropics have a special contribution to make to biological knowledge in respect of phenomena that are unknown in temperate regions. Almost all our understanding of plant physiology and ecology has been gained from studies of temperate plants. It is not too much to say that when an equivalent amount of research has been done in the tropics it will be necessary to re-write the textbooks, including indeed this one.

Appendix: Conversion diagrams for metric and English measurements.

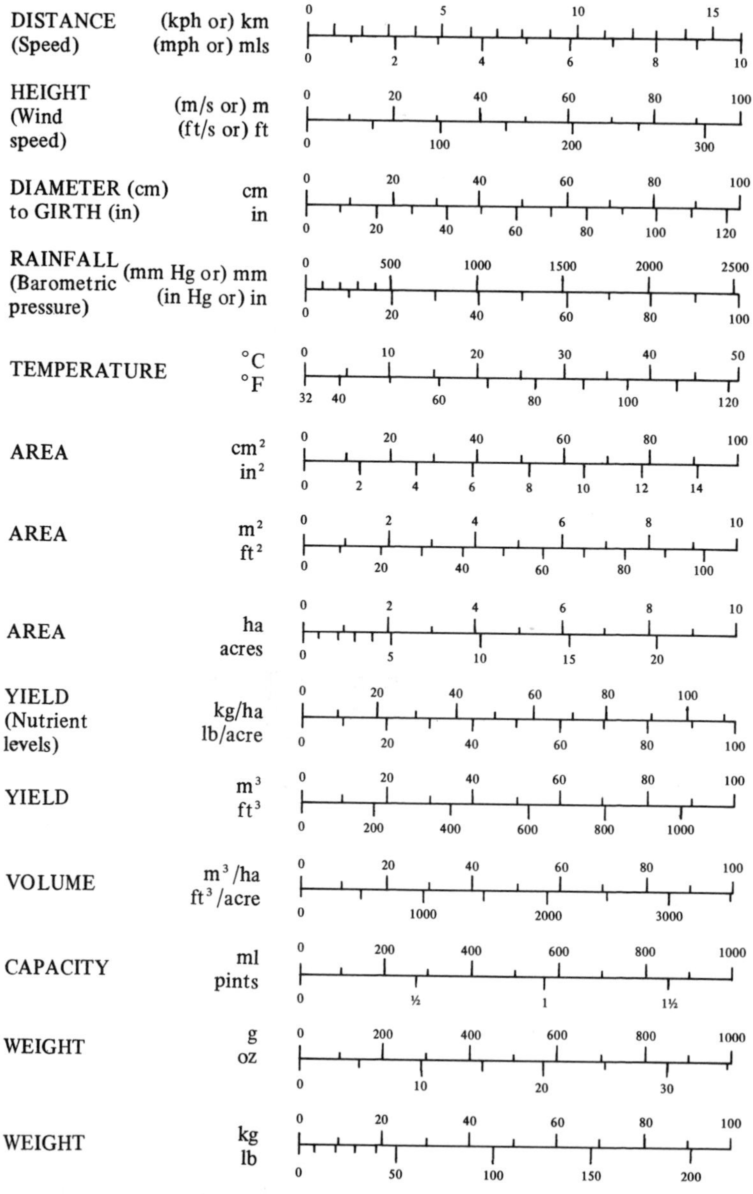

WEIGHT One metric tonne is almost exactly (98·4 per cent of) an English ton. All these scales, except that for temperature, can be multiplied (or divided) by 10 or any other convenient factor.

Plate 1. Freshly-cut vertical profile through a tropical evergreen seasonal forest near Kade, Ghana, showing layering, emergents and climbers of a virgin stand. Compare also the biomass (standing crop) of the forest with the maize farm in foreground. July 1967.

Plate 2. Various crown shapes of tropical forest emergents. From left to right: *Tarrietia utilis, Pentadesma butyracea* and *Lophira alata.* December/January.

Plate 3. Interior of an ombrophilous forest in the Ankasa Forest Reserve, Ghana, showing *Pycnocoma macrophylla*, a typical unbranched pigmy tree.

Plate 4A. Interior of an African ombrophilous forest showing the relative abundance of suppressed seedlings, saplings and pigmy trees. Note also the rotting fallen logs on the forest floor.

4B. Characteristic view of the lower layers of an ombrophilous forest showing absence of well-developed herbaceous synusia. Among numerous smaller trees and the loops of lianes is the base of a single emergent of *Dialium aubrévillei* with buttresses.

Plate 5. Crown of a giant emergent, deciduous, heavily laden with epiphytes, such as *Oleandra*, *Drynaria*, *Peperomia*, etc. May 1967.

Plate 6. Rapid regrowth of thicket on forest margin, composed mainly of *Musanga cecropioides* and *Lophira alata* (lanceolate leaves). Approx. 3 years old since clearing; the large trees with heavy, dense foliage are *Tarrietia utilis.*

Plate 7A. Drip-tips of leaves in the undergrowth of the ombrophilous forest. Note the partial destruction of leaves by insects.

7B. An example of the frequently occurring leaf-joints (pulvini) in a tropical tree (*Khaya ivorensis*). Note the 'wrinkles' suggestive of movements of the leaf.

Plate 8A. Base of stem of large woody climber (*Neuropeltis prevosteoides*) showing unusual contortion due to abnormal secondary thickening and looping.

8B. Fruits borne directly on the main stem resulting from cauliflory (in the narrow sense). The single seed is projected explosively from the capsule in this pigmy tree *Allexis cauliflora.*

Plate 9A. Characteristic appearance of the leaf surface of a forest herb (*Cephaelis* sp.) in the dim light near the floor.

9B. Broad leaves and inflorescence of a tropical forest sedge, *Mapania* sp.

Plate 10. Stilt-roots of *Xylopia staudtii*, growing on a mesic site in the Atewa Range, Ghana. On impeded soils, this species may develop stilted peg-roots (see p. 59).

Plate 11A. Epiphyllous synusia of lichens, mosses, liverworts and algae on the surface of a living leaf.

11B. Characteristic fruit body of a Basidiomycete *Dictyophora phalloidea* in the litter of a rain forest.

Plate 12A. Abandoned farm with the remains of a crop of cassava and plantain in the Atewa Range, Ghana. May 1967.

12B. Considerable erosion occurring within three years of opening a forest on a steeper slope on ferralitic soil. A *Lophira* seedling is invading the devastated area.

B

Plate 13A. Adventitious roots on the lower portion of the trunk of *Afrosersalisia afzelii*. This type of aerial root seldom develops into thick stilts reaching the soil.

13B. Peg-roots of *Anthocleista nobilis* represent a frequent type of pneumorhizae in freshwater swampy soils. Because of their colour and small size they can easily be overlooked.

Plate 14A. 'Slash' of *Celtis mildbraedii* showing conspicuous tangential banding which is important for the identification of this emergent; the dark zones are bright red-brown.

14B. Large fruts of the cauliflorous tree *Omphalocarpum ahia*. These heavy fruits easily fall through the canopy to the forest floor.

Plate 15A. Broken limb from the upper tree layer, lying on the forest floor. Shows epiphytes of microhabitat no. 2 (see Fig. 4.2), the dominant species being *Peperomia fernandopoiana.*

15B. Tropical ombrophilous mountain forest in Ghana, with numerous epiphytes including hanging clusters of *Orthostichidium perpinnatum.* Mr A. A. Enti, formerly keeper of the herbarium at the University of Ghana.

Plate 16A. Base of a tree bole (zone 5 in Fig. 4.2), richly covered with epiphytic mosses, filmy ferns and a species of *Begonia*.

16B. *Dissotis entii*, in the Melastomataceae, a characteristic family of the undergrowth in the African ombrophilous forest.

Plate 17A. Strangling fig on an oil-palm at Rokupr, Sierra Leone. (See also Fig. 4.3).

17B. *Cedrela odorata* of a provenance from Misiones province, Argentina, growing in the International provenance trial at Sapoba, Nigeria. The tree is completely dormant, and also shows the point at which it was long dormant earlier. Planted May 1969; photographed Dec. 8, 1970. Height less than 1 m. By contrast, a provenance from British Honduras had plot mean heights five times greater, and showed little or no signs of other than temporary dormancy, past or present. The biggest tree (18 months old) was 10 m tall and had a diameter of 10 cm at 1·3 m from the ground.

Plate 18. A young silk cotton tree (*Ceiba pentandra*) in southern Ghana, showing different phases: lower left – newly flushing leaves; centre – mature leaves; upper and right – deciduous phase, with both ripening and shedding fruits. Early Feb. 1967.

Plate 19A. *Ceiba pentandra* seedlings showing effects of day-length. LD – 17·2 h; GLD – 13·2 h; SD – 9·2 h. The two SD plants on the left are dormant; the other two have re-flushed after earlier temporary dormancy. All with 26°C night temperature.

19B. The original Cotton Tree, well over 200 years old, which stands in a prominent position near the centre of Freetown, Sierra Leone, and under which many slaves were given their freedom.

Plate 20. Effect of day-length and night temperature on the shoot growth of A – *Terminalia ivorensis* and B – *Triplochiton scleroxylon* seedlings. Rows 1 and 2 from the left received 20°C nights; 3 and 4, 30°C nights: day temperatures were similar for all treatments. Rows 1 and 3 were grown under 11 h days; 2 and 4 had 14½ h days. Photos after two months' treatment. Note the interactions:

A – much faster shoot extension in row 4;

B – much slower leaf expansion in row 1.

Plate 21. Effect of day-length on terminal bud dormancy of *Cedrela odorata* seedlings. A – tree under long-days, showing actively growing shoot apex. B – apex dormant under short-days.

Plate 22A. View looking directly upwards towards the canopy of an early secondary stage forest in the Loma Mts., Sierra Leone. The trees on the middle left and centre are *Fagara macrophylla*: at the top left and centre are trees of *Musanga cecropioides*. June 1962.

22B. Leaf-exchanging habit in *Terminalia catappa.* The old, senescent leaves are here falling simultaneously with the active flushing of new leaves.

Plate 23A. Young *Terminalia ivorensis* plantation, showing the 'pagoda' habit. In some trees, growth is taking place in the new, sub-terminal branches, while other individuals are making very rapid extension of leading shoots.

23B. Top of the large *Terminalia ivorensis* tree shown in Plate 25A. The sub-terminal branches have far outstripped the growth of the main apex ('lateral dominance'). Note the early stages in the flushing and restarting of leader growth, and the fact that there has been considerable leaf-fall in the basal portions of the branches.

Plate 24. Contrasting growth habits of juvenile and mature tissue. A – *Terminalia ivorensis*, seedling on the left; mature graft on the right. B – *Cedrela odorata*, four seedlings on the left, with vertical shoots growing up after pruning back; four mature grafts on the right, made on seedlings from the same batch, showing variable orientation of shoot systems. Budding of grafts carried out by Mr N. Jones. Several years after these mature trees had been planted out, the growth habit was still different from trees of seedling origin.

Plate 25A. Eight seedlings of *Terminalia ivorensis*, showing the powerful influence of the environment. They are all 5 years old and very 'pot-bound', except the one on the left, which had rooted through the hole at the bottom of the pot, and then cracked it into pieces. It then proceeded to flourish, presumably favoured by rooting freely into the soil, and by seepage of water and nutrients from the greenhouse. Note that it has entered the deciduous phase, while the other seven were all evergreen, and that bud-break of the leading shoot has just occurred. University of Ghana, Legon. May 1968.

Plate 25B. View northwards from the side of a steep rocky hill near Ondo, Nigeria. In the foreground are *Hildegardia barteri* seedlings growing in a small crack in the granite; oil-palms and secondary forest in the background, with 'harmattan' haze. Dec. 9, 1970.

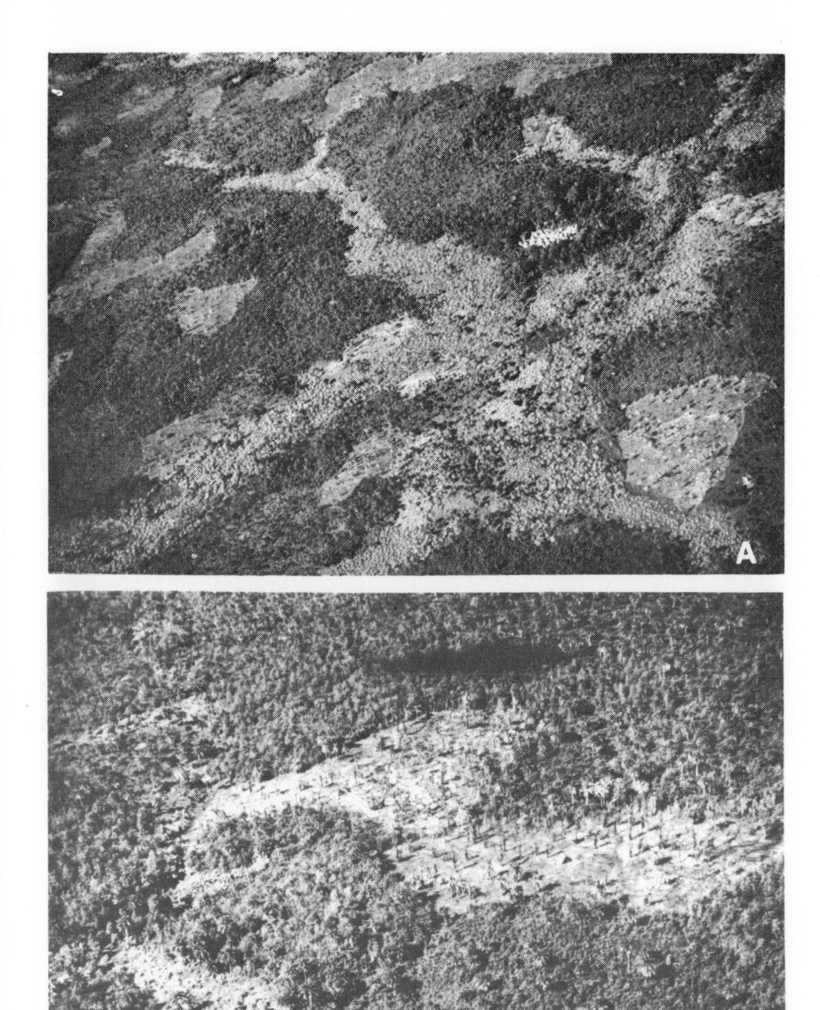

Plate 26. Air photographs of farm-bush near Bo, Sierra Leone. A – the paler X-shaped area is a swamp with *Raphia* palms at the bottom of the catena; the village and most of the current year's farms (pale patches) are on the mesic sites. The bush-fallow period appears to be only about 5 years, and the only large trees are a few near to the village. B – closer view of a current year's farm, burnt a few months previously. Note that the oil-palms have been left scattered over the farm: they are fairly fire-resistant, and are amongst the tallest trees in the surrounding farm-bush. June 1962.

Plate 27. *Eurychobe rothschildiana*, an epiphyte in the West African forests.

Plate 28. An unusual species of the African forests, *Microcoelia caespitosa.* This epiphytic orchid has an extremely reduced shoot system (apart from the inflorescences); thick, green roots form the major portion of the plant and its organs of photosynthesis.

References

ADDICOTT, F. T. and LYNCH, R. S. (1955). 'Physiology of abscission', *A. Rev. Pl. Physiol.*, **6**, 211–38.

AHN, P. (1961). 'Soils of the Lower Tano Basin, south-western Ghana,' *Soil & Land-Use Survey*, Mem. no. 2. Govt. Printer, Accra.

AHN, P. (1970). *West African Soils*, Oxford University Press.

ALLEE, W. C. (1926). 'Measurement of environmental factors in the tropical rain-forest of Panama', *Ecology*, **7**, 273–302.

ALVIM, P. de T. (1960). 'Moisture stress as a requirement for flowering of coffee', *Science*, **132**, 354.

ALVIM, P. de T. (1964). 'Tree growth periodicity in tropical climates', in *Formation of Wood in Forest Trees*, ed. M. H. Zimmermann, pp. 479–95, New York. Academic Press.

ALVIM, P. de T. (1967). 'Ecophysiology of the cocoa tree', *Conf. int. sur. les recherches agronomiques cacaoyères, Abidjan 1965,* pp. 23–35, Paris.

ALVIM, P. de T. and GRANGIER Jr., A. (1965). Influência do fotoperiodo no lançamento do cacaueiro', *Centr. Pesq. Cacau, Itabuna, Brasil, Ann. Rep.*, 1964, p. 24.

ANONYMOUS (1960). 'Records of forest plantation growth in Mexico, the West Indies, and Central and South America', *Caribb. Forester*, **21** (supplement).

ASHTON, P. S. (1964). 'Ecological studies in the mixed dipterocarp forests of Brunei State', *Oxford Forestry Memoires*, **25**, 1–75.

AUBRÉVILLE, A. (1938). 'La forêt coloniale: les forêts de l'Afrique occidentale française', *Ann. Acad. Sci. colon., Paris*, **9**, 1–245.

AUBRÉVILLE, A. (1949). *Contribution à la paléohistoire des forêts de l'Afrique tropicale*, Paris.

AUSTIN, M. P. and GREIG-SMITH, P. (1968). 'The application of quantitative methods to vegetation survey. II. Some methodological problems of data from rain forest', *J. Ecol.*, **56**, 827–44.

BAKER, H. G. (1965). 'The evolution of the cultivated Kapok Tree: a probable West African product'. In *Ecology and Economic Development in Tropical Africa*, ed. D. Brokensha, pp. 185–216. Berkeley. University of California.

BAKER, H. G. and HARRIS, B. J. (1959). 'Bat-pollination of the silk-cotton tree, *Ceiba pentandra* (L.)Gaertn. (sensu latu) in Ghana', *Jl. W. Afr. Sci. Ass.*, **5**, 1–9.

BAKER, J. R. and BAKER, I. (1936). 'The seasons in a tropical rain-forest (New Hebrides). Part 2. Botany', *J. Linn. Soc. (Zool)*, **39**, 507–19.

BARUA, D. N. (1969). Seasonal dormancy in tea (*Camellia sinensis* L.)', *Nature, Lond.*, **224** , 514.

BEARD, J. S. (1942). 'The use of the term "deciduous" as applied to forest types in Trinidad', *Emp. For. J.*, **21**, 12–17.

BEARD, J. S. (1945). 'The progress of plant succession on the Soufrière of St. Vincent', *J. Ecol.*, **33**, 1–9.

BECKING, W. (1960). 'A summary of information on *Aucoumea klaineana*', *For. Abstr.*, **21**, 1–6, 163–72.

BOND, T. E. T. (1942). 'Studies in the vegetative growth and anatomy of the tea plant (*Camellia thea* Link.) with special reference to the phloem. I. The flush shoot', *Ann. Bot.*, **6**, 607–29.

BOND, T. E. T. (1945). 'Studies in the vegetative growth and anatomy of the tea plant (*Camellia thea* Link.) with special reference to the phloem. II. Further analysis of flushing behaviour', *Ann Bot.*, **9**, 183–216.

BOOTH, A. H. (1956). 'The distribution of primates in the Gold Coast', *Jl. W. Afr. Sci. Ass.*, **2**, 122–33.

BOYER, J. (1969). 'Étude expérimentale des effets du régime d'humidité du sol sur la croissance végétative, la floraison et la fructification des caféiers robusta', *Café, cacao, thé*, **13**, 187–200.

BRADLEY, J. W. (1922). 'A plantation of remarkable growth', *Indian Forester*, **48**, 637–40.

BRAMMER, H. (1962). 'Soils', in: *Agriculture and Land Use in Ghana*, ed. J. B. Wills, pp. 88–126. Oxford University Press.

BRAY, J. R. and GORHAM, E. (1964). 'Litter production in forests of the world', in: *Advances in Ecological Research*, vol. 2, ed. J. B. Cragg, pp. 101–57. London. Academic Press.

BRONCHART, R. (1963). 'Recherches sur le développement de *Geophila renaris* De Wild. et Th.Dur. dans les conditions écologiques d'un sous-bois forestier équatorial. Influence sur la mise à fleurs d'une perte en eau disponible du sol', *Mém. Soc. Roy. Sci. Liège*, sér. 5, **8** (2), 1–181.

BROWNING, G. (1971). *The Hormonal Regulation of Flowering and Cropping in Coffee arabica L.* Ph.D. Thesis, University of Bristol.

BUDOWSKI, G. (1965). 'Distribution of tropical American rain forest species in the light of successional processes', *Turrialba*, **15**, 40–2.

BÜNNING, E. (1947). *In den Wäldern Nordsumatras*, Bonn. F. Dümmlers.

BÜNNING, E, (1948). 'Studien über Photoperiodizität in den Tropen', in: *Venalisation and photoperiodism*, eds. A. E. Murneek and R. O. White, pp. 161–66. Chronica Botanica.

BÜNNING, E. (1956). *Der tropische Regenwald*, Berlin. Springer.

BUNTING, B. T. (1967). *The Geography of Soil*, 2nd edn. London. Hutchinson.

BURTT DAVY, J. (1938). 'The classification of tropical woody vegetation types', *Imp. For. Inst., Oxford*, Paper no. 13.

CACHAN, P. (1963). Signification écologique des variations microclimatiques verticales dans la forêt sempervirente de Basse Côte d'Ivoire', *Ann. Fac. Sci. Dakar*, **8**, 89–155.

CACHAN, P. and DUVAL, J. (1963). 'Variations microclimatiques verticales et saisonnières dans la forêt sempervirente de Basse Côte d'Ivoire', *Ann. Fac. Sci. Dakar*, **8**, 5–87.

CALLAHAM, R. Z. (1962). Geographic variability in growth of forest trees', in: *Tree Growth*, ed. T. T. Kozlowski, pp. 311–25. New York. Ronald Press.

CARTER, G. S. (1934). 'Reports of the Cambridge Expedition to British Guiana, 1933. Illumination in the rain forest at ground level', *J. Linn. Soc. (Zool)*, **38**, 579–89.

COOMBE, D. E. and HADFIELD, W. (1962). 'An analysis of the growth of *Musanga cecropioides*', *J. Ecol.*, **50**, 221–34.

CORNER, E. J. H. (1940). *Wayside Trees of Malaya*, vol. 1. Singapore. Govt. Printer.

CORNER, E. J. H. (1954). 'The evolution of tropical forest', in: *Evolution as a Process*, eds. J. S. Huxley, A. C. Hardy and E. B. Ford, pp. 34–46. London. Allen & Unwin.

COSTER, C. (1923). Lauberneuerung und andere periodische Lebensprozesse in dem trockenen Monsungebiet Ost-Javas', *Annls. Jard. bot. Buitenz.*, **33**, 117–89.

COSTER, C. (1926). 'Periodische Blüterscheinungen in den Tropen', *Annls. Jard. bot. Buitenz.*, **35**, 125–62.

COSTER, C. (1927). 'Die tägliche Schwankungen des Längenzuwachses in den Tropen', *Recl. Trav. bot. néer.*, **24**, 257–305.

COSTER, C. (1927–28). 'Zur Anatomie und Physiologie der Zuwachszonen und Jahresringbildung in den Tropen', *Annls. Jard. bot. Buitenz.*, **37**, 49–160; **38**, 1–114.

COSTER, C. (1932). 'Wortelstudiën in de Tropen. I. De jeugdontwikkeling van het wortelstelsel van een zeventigtal boomen en groenbemesters', *Tectona*, **25**, 828–72.

COSTER, C. (1933). 'Wortelstudiën in de Tropen. III. De zuurstof behoefte van het wortelstelsel', *Tectona*, **26**, 450–97.

CRITCHFIELD, H. J. (1966). *General Climatology*. Prentice-Hall. New Jersey.

DAMPTEY, H. B. (1964). *Studies of Apical Dominance in Woody Plants*. MSc Thesis, University of Ghana.

DAWKINS, H. C. (1964). 'Productivity of tropical forests and their ultimate value to man', *IUCN Publ.*, new series, **4**, 178–82.

DAWKINS, H. C. (1967). *The Production of Tropical High-forest Trees and Their Reaction to Controllable Environment*, PhD Thesis, Oxford University Press.

DOBZHANSKY, T. (1950). 'Evolution in the tropics', *Amer. Scientist*, **38**, 209–21.

DUMONT, R. (1966). *False Start in Africa*. Translated by P. N. Ott. Deutsch.

ELLENBERG, H. (1959a). 'Über den Wasserhaushalt tropischer Nebeloasen in der Küstenwüste Perus', *Ber. geobot. ForschInst. Rübel*, **1958**, 47–74.

ELLENBERG, H. (1959b). 'Typen tropischer Urwälder in Peru', *Schweiz. Z. f. Forstwesen*, **3**, 1–19.

ELLENBERG, H. and MUELLER-DOMBOIS, D. (1967). 'Tentative physiognomic-ecological classification of plant formations of the earth', *Ber. geobot. ForschInst. Rübel*, **37**, 21–55.

ENTI, A. A. (1968). 'Distribution and ecology of *Hildegardia barteri* (Mast.)Kosterm', *Bull. de l'IFAN*, (sér. A.), **30**, 881–95.

EVANS, G. C. (1956). 'An area survey method of investigating the distribution of light intensity in woodlands, with particular reference to sunflecks, including an analysis of data from rain forest in Southern Nigeria', *J. Ecol.*, **44**, 391–428.

ÉVRARD, C. (1968). 'Recherches écologiques sur le peuplement forestier des sols hydromorphes de la cuvette centrale congolaise', *I.N.É.A.C., Brussels*, publ. no. **110**, 1–295.

FAEGRI, K. and PIJL, van der L. (1966). *The Principles of Pollination Ecology*. Toronto. Pergamon Press.

FEDEROV, A. A. (1966). 'The structure of the tropical rain forest and speciation in the humid tropics', *J. Ecol.*, **54**, 1–11.

FOXWORTHY, F. W. (1927). *Commercial Timber Trees of the Malay Peninsula*. Malay. For. Rec., 3.

FREISE, F. (1936). 'Das Binnenklima von Urwäldern in subtropischen Brasilien', *Petermann's Mitt.*, **82** 301–7.

FUNKE, G. L. (1929, 1931). 'On the biology and anatomy of some tropical leaf-joints. Parts I and II', *Annls. Jard. bot. Buitenz.*, **40**, 45–74; **41**, 33–64.

FURR, J. R., COOPER, W. C. and REECE, P. C. (1947). 'An investigation of flower formation in adult and juvenile citrus trees', *Am. J. Bot.*, **34**, 1–8.

GERMAIN, R. (1963). 'Les biotopes alluvionnaires herbeux et les savanes intercalaires du Congo équatoriale', *Acad. roy. Sci. d'outre-mer, Cl. Sci. nat. et médic.*, N.S. **15** (4), 1–399.

GERMAIN, R. and ÉVRARD, C. (1956). 'Étude écologique et phytosociologique de la forêt à *Brachystegia laurentii*', *I.N.É.A.C., Brussels*, publ. no. **67**, 1–105.

GIBBS, D. G. and LESTON, D. (1970). 'Insect phenology in a forest farm locality in West Africa', *J. Appl. Ecol.*, **7**, 519–48.

GOEBEL, K. (1928). *Organographie der Pflanzeh*, Teil I. Jena. G. Fischer.

GOUROU, P. (1953). *The Tropical World*. Translated by E. D. Laborde. London. Longmans.

GREENWOOD, M. and POSNETTE, A. F. (1949). 'The growth flushes of cocoa', *J. hort. Sci.*, **25**, 164–74.

GREIG-SMITH, P., AUSTIN, M. P. and WHITMORE, T. C. (1967). 'The application of quantitative methods to vegetation survey. Part I. Association-analysis and principal component ordination of rain forest', *J. Ecol.*, **55**, 483–503.

GUILLAUMET, J. L. (1967). *Recherches sur la végétation et la flore de la région du Bas-Cavally (Côte-d'Ivoire).* Paris. O.R.S.T.O.M.

HACKETT, W. P. and SACHS, R. M. (1966). 'Flowering in Bougainvillea "San Diego" ', *Proc. Am. Soc. hort. Sci.*, **88**, 606–12.

HALLÉ, F. and OLDEMAN, R. A. A. (1970). *Essai sur l'architecture et la dynamique de croissance des arbres tropicaux*. Paris. Maçon et Cie.

HALLE, F. and MARTIN, R. (1968). 'Study of the growth rhythm in *Hevea brasiliensis* (Euphorbiaceae-Crotonoideae), *Adansonia, Paris*, **8**, 470–503.

HARDON, H. J. (1937). 'Padang soil, an example of podzol in the tropical lowlands', *Verh. K. Akad. Wet.*, **40**, 530–38.

HARDY, F. (1958). 'The effects of air temperature on growth and production in cacao', *Cacao*, **3** (17), 1–14.

HARDY, F. (1964). 'Soils and ecology of the cacao belt of Ecuador', *Cacao*, **9**, 1–23.

HOLDSWORTH, M. (1959). 'The effects of daylength on the movements of pulvinate leaves', *New Phytol*, **58**, 29–45.

HOLDSWORTH, M. (1961). 'The flowering of rain flowers', *Jl. W. Afr. Sci. Ass.*, **7**, 28–36.

HOLTTUM, R. E. (1940). 'Periodic leaf-change and flowering of trees in Singapore', *Gdns' Bull., Singapore*, **11**, 119–75.

HOLTTUM, R. E. (1953): 'Evolutionary trends in an equatorial climate'. *Symp. Soc. exp. Biol.*, **7**, 159–73.

HOPKINS, B. (1970). 'Vegetation of the Olokemeji Forest Reserve, Nigeria. VI. The plants on the forest site, with special reference to their seasonal growth', *J. Ecol.*, **58**, 765–93.

HUECK, K. (1966). *Die Wälder Südamerikas*. Stuttgart. G. Fischer.

HUGHES, A. P. (1966). 'The importance of light compared with other factors affecting plant growth', in: *Light as an Ecological Factor*, eds. R. Bainbridge, G. C.Evans and O. Rackham, pp. 121–46. Oxford. Blackwell.

HUMMEL, F. C. (1946). 'The formation of growth rings in *Entandrophragma macrophyllum* A. Chev. and *Khaya grandifoliola* C.DC., *Emp. For. Rev.*, **25**, 103–7.

HUSSEY, G. (1965). 'Growth and development in the young tomato. III. The effect of night and day temperatures on vegetative growth', *J. exp. Bot.*, **48**, 373–85.

HUTCHINSON, J., DALZIEL, J. M. and KEAY, R. W. J. (1954). *Flora of West Tropical Africa*, vol. 1, part 1, 2nd edn. London. Crown Agents.

HUXLEY, P. A. (1970). 'Some aspects of the physiology of arabica coffee – the central problem and need for a synthesis', in: *Physiology of Tree Crops*, eds. L. C. Luckwill and C. V. Cutting, pp. 255–68. New York. Academic Press.

IRVINE, F. R. (1961). *Woody Plants of Ghana.* Oxford. University Press.

JENÍK, J. (1967). 'Root adaptations in West African trees'. *J. Linn. Soc. (Bot)*, **60**, 25–9.

JENÍK, J. (1969). 'The life-form of *Scaphopetalum amoenum* A. Chev.', *Preslia*, **41**, 109–12.

JENÍK, J. (1970a). The pneumatophores of *Voacanga thouarsii* Roem. et Schult. (Apocynaceae)', *Bull. de l'IFAN*, **32** (A), 986–94.

JENÍK, J. (1970b). 'Root system of tropical trees 5. The peg-roots and the pneumathodes of *Laguncularia racemosa* Gaertn.', *Preslia*, **42**, 105–13.

JENÍK, J. (1970c). 'Root system of tropical trees 4. The stilted peg-roots of *Xylopia staudtii* Eng. et Diels', *Preslia*, **42**, 25–32.

JENÍK, J. (1971a). 'Root system of tropical trees 6. The aerial roots of *Entandrophragma angolense* (Welw.)C.DC.', *Preslia*, **43**, 1–4.

JENÍK, J. (1971b). 'Root system of tropical trees 7. The facultative peg-roots of *Anthocleista nobilis* G.Don.', *Preslia*, **43**, 97–104.

JENÍK, J. (1971c). 'Root structure and underground biomass in equatorial forests,' in: *La productivité des écosystèmes forestiers*, ed. P. Duvigneaud, pp. 323–31. Paris, UNESCO.

JENÍK, J. (1973). 'Root system of tropical trees. 8. Stilt-roots and allied adaptations', *Preslia*, **45**, 250–64.

JENIK, J. and ENTI, A. A. (1969). 'Discontinuous distribution of *Allexis cauliflora* (Oliv.) L.Pierre in Equatorial Africa', *Novit. bot. Inst. bot. Univ. carol. prag.*, 1968, 67–71.

JENÍK, J. and HALL, J. B. (1966). 'The ecological effects of the harmattan wind in Djebobo Massif, Togo', *J. Ecol.*, **54**, 767–79.

JENÍK, J. and HARRIS, B. J. (1969). 'Root-spines and spine-roots in dicotyledonous trees of Tropical Africa', *Öst. bot. Z..*, **117**, 128–38.

JENÍK, J. and MENSAH, K. O. A. (1967). 'Root system of tropical trees. 1. Ectotrophic mycorrhizae of *Afzelia africana* Sm.', *Preslia*, **39**, 59–65.

JONES, E. W. (1955. 1956). 'Ecological studies on the rain forest of Southern Nigeria. IV. The plateau forest of the Okumu Forest Reserve. Part I and II', *J. Ecol.*, **43**, 564–94; **44**, 83–117.

JONES, N. (1967). Private communication.

KEAY, R. W. (1960). 'Seeds in forest soils', *Niger. For. Inf. Bull*, n.s. **4**, 1–12.

KIRA, T. and SHIDEI, T. (1967). 'Primary production and turnover of organic matter in different forest ecosystems of the western Pacific', *Jap. J. Ecol.*, **17**, 70–87.

KLEBS, G. (1926). 'Über periodisch wachsende tropische Baumarten', *Sber. heidelb. Akad. Wiss., Math.–Naturwiss. Kl.*, **2**, 1–31.

KOOPER, W. J. C. (1927). 'Sociological and ecological studies on the tropical weed-vegetation of Pasurvan (the island of Java)', *Recl. Trav. bot. néer.*, **24**, 1–256.

KORIBA, K. (1958). 'On the periodicity of tree-growth in the tropics, with reference to the mode of branching, the leaf-fall, and the formation of the resting bud', *Gdns' Bull., Singapore*, **17**, 11–81.

KOZLOWSKI, T. T. and KELLER, T. (1966). 'Food relations of woody plants', *Bot. Rev.*, **32**, 293–382.

KRAMER, P. J. and KOZLOWSKI, T. T. (1960). *Physiology of Trees*. New York. McGraw-Hill.

KREKULE, J. (1969). 'Cherelle-wilting in cocoa induced by T.I.B.A.', *Trop. Agric, Trinidad*, **46**, 69–72.

KREKULE, J. (1972). Private communication.

KWAKWA, R. S. (1964). *The Effects of Temperature and Day-length on Growth and Flowering in Woody Plants.* MSc Thesis, University of Ghana.

LAMB, A. F. A. and NTIMA, O. O. (1971). *Fast Growing Timber Trees of the Lowland Tropics*. No. 5. *Terminalia ivorensis*. Oxford. Comm. For. Inst.

LAMPRECHT, H. (1961). *Tropenwälder und tropische Waldwirtschaft.* Beih. schweiz. Z. Forstw. No. 32, Zürich.

LANGDALE-BROWN, I., OSMASTON, H. A. and WILSON, J. G. (1964). *The Vegetation of Uganda, and its Bearing on Land-Use.* Entebbe. Uganda Govt.

LARSON, P. R. (1964). 'Some indirect effects of environment on wood formation', in: *The Formation of Wood in Forest Trees*, ed. M. H. Zimmermann, pp. 345–65. New York. Academic Press.

LAWSON, G. W., ARMSTRONG-MENSAH, K. O. and HALL, J. B. (1970). 'A catena in tropical moist semi-deciduous forest near Kade, Ghana', *J. Ecol.*, **58**, 371–98.

LEAKEY, L. S. B. (1964). 'Prehistoric man in the tropical environment', in: *The Ecology of Man in the Tropical Environment*, p. 24–29. Morges. I.U.C.N.

LEBRUN, J. (1947). 'La végétation de la plaine alluviale au sud du Lac Édouard. Expl. Parc Nat. Albert, mission J.Lebrun (1937–38)', *Inst. Parcs Nat. Congo belge*, **1**, 1–800.

LEBRUN, J. and GILBERT, G. (1954). 'Une classification écologique des forêts du Congo', *I.N.É.A.C., Brussels*, publ. no. **63**, 1–89.

LEMÉE, G. (1956). 'Recherches écophysiologiques sur les cacaoyers', *Revue gén. Bot.*, **63**, 41–94.

LETOUZEY, R. (1969, 1970). *Manuel de botanique forestière*, vols. 1 and 2. Nogent s/Marne. Centre tech. forest trop.

LONGMAN, K. A. (1966). 'Effects of the length of the day on growth of West African trees', *Jl. W. Afr. Sci. Ass.*, **11**, 3–10.

LONGMAN, K. A. (1969). 'The dormancy and survival of plants in the humid tropics', *Symp. Soc. exp. Biol.*, **23**, 471–88.

LONGMAN, K. A. (1972). 'Environmental control of shoot growth in tropical trees', in: *Essays in Forest Meteorology: an Aberystwyth Symposium*, ed. J. A. Taylor, pp. 157–67, Aberystwyth.

LONGMAN, S. J. (1964). Private communication.

LOWE, R. G. (1968). 'Periodicity of a tropical rain forest tree, *Triplochiton scleroxylon* K. Schum.', *Comm. For. Rev.*, **47**, 150–63.

LYR, H. and HOFFMANN, G. (1967). 'Growth rates and growth periodicity of tree roots', *Int. Rev. For. Res.*, **2**, 181–236. New York. Academic Press.

MALLIK, P. C. (1951). 'Inducing flowering in mango by ringing the bark', *Ind. J. Hort.*, **8**, 1–10.

MARGALEF, R. (1968). *Perspectives in Ecological Theory*. Chicago. University Press.

MARRERO, J. (1942). 'A seed storage study of Maga', *Caribb. Forester*, **3**, 173–84.

MARRERO, J. (1943). 'A seed storage study of some tropical hardwoods', *Caribb. Forester*, **4**, 99–106.

McKELVIE, A. D. (1954). 'Root studies on seedlings', *W. Afr. Cocoa Res. Stn., Tafo, Ghana, Ann. Rep.*, 1953–54, p. 24–25.

McKELVIE, A. D. (1956). 'Cherelle wilt of Cacao. I. Pod development and its relation to wilt', *J. Exp. Bot.*, **7**, 252–63.

McKELVIE, A. D. (1958). 'Root studies', *W. Afr. Cocoa Res. Stn., Tafo, Ghana, Ann. Rep.* 1956–57, p. 57–58.

MEIKLEJOHN, J. (1962). 'Microbiology of the nitrogen cycle in some Ghana soils', *Emp. J. exp. Agric.*, **30**, 115–62.

MENSAH, K. O. A. and JENÍK, J. (1968). 'Root system of tropical trees. 2. Features of the root system of iroko (*Chlorophora excelsa* Benth. et Hook.)', *Preslia*, **40**, 21–27.

MES, M. G. (1956–57). 'Studies on the flowering of *Coffea arabica*', *Port. Acta biol.*, **4**, 328–41.

MOREL, G. (1960). 'Physiologie du cambium', in *Colloque de xylologie, 1959*, pp. 50–60, *Mém. Soc. bot. Fr.*, 1960.

MORRISON, C. G. T., HOYLE, A. C. and HOPE-SIMPSON, J. F. (1948). 'Tropical soil-vegetation catenas and mosaics', *J. Ecol.*, **36**, 1–84.

MURRAY, D. B. (1964). 'Morphological effects of temperature on the growth of *Theobroma cacao*', *Nature, Lond.*, **202**, 1134.

MURRAY, D. B. (1966). 'Soil moisture regimes', *Ann. Rep. Cacao Res., Trinidad*, 1965, pp. 34–39.

MURRAY D. B. and NICHOLS, R. (1966). 'Light, shade and growth in some tropical plants', in: *Light as an Ecological Factor*, eds. R. Bainbridge, G. C. Evans and O. Rackham, pp. 249–63, Oxford. Blackwell.

MURRAY, D. B. and SALE, P. J. M. (1966). 'Report on plant physiology', *Ann. Rep. Cacao Res., Trinidad*, 1965, pp. 30–4.

MURRAY, D. B. and SALE, P. J. M. (1967). 'Growth studies on cacao in controlled environment rooms', *Conf. int. sur les recherches agronomiques cacaoyeres*, Abidjan 1965, pp. 57–63. Paris.

NAUNDORF, G. (1954). 'Contribution a la fisiologia de la floracion en cacao. Existencia de hormonas de floracion', *Cacao en Colombia*, **3**, 29–34.

NG, F. S. P. (1966). 'Age at first flowering in dipterocarps', *Malay. Forester*, **29**, 290–95.

NITSCH, J. P. (1957). 'Photoperiodism in woody plants', *Proc. Am. Soc. hort. Sci.*, **70**, 526–44.

NJOKU, E. (1963). 'Seasonal periodicity in the growth and development of some forest trees in Nigeria', *J. Ecol.*, **51**, 617–24.

NJOKU, E. (1964). 'Seasonal periodicity in the growth and development of some forest trees in Nigeria. Part 2. Observations on seedlings', *J. Ecol.*, **52**, 19–26.

NOEL, A. R. A. (1970). 'The girdled tree', *Bot. Rev.*, **36**, 162–95.

NYE, P. H. and GREENLAND, D. J. (1960). *The Soil under Shifting Cultivation.* Commonwealth Agricultural Bureaux.

OBATON, M. (1960). 'Les lianes ligneuses à structure anomale des forêts denses d'Afrique occidentale, *Annls. Sci. nat.*, sér. A, **12**, 1–220.

OGURA, Y. (1940). 'On the types of abnormal roots in mangroves and swampy plants'. *Bot. Mag., Tokyo*, **54**, 389–404.

OKALI, D. U. U. (1971). 'Rates of dry-matter production in some tropical forest-tree seedlings,' *Ann. Bot.*, **35**, 87–97.

OLATOYE, S. T. (1968). *Seed Storage Problems in Nigeria.* Paper at 9th B.C.F.C. New Delhi, India.

OLATOYE, S. T. Private communication.

OWEN, D. F. (1966). *Animal Ecology in Tropical Africa.* Edinburgh.

OWEN, G. (1951). 'A provisional classification of Malayan soils', *J. Soil Sci.*, **2**, 20–42.

PEYRONEL, B. and FASSI, B. (1957). 'Micorrize ectotrofiche in una cesalpiniacea del Congo Belga', *Atti Accad. Sci., Torino*, **91**, 569–76.

PHILLIPS, J. (1959). *Agriculture and Ecology in Africa.* London. Faber & Faber.

PIRINGER, A. A. and BORTHWICK, H. A. (1955). 'Photoperiodic responses of coffee', *Turrialba*, **5**, 72–77.

PIRINGER, A. A. and DOWNS, R. J. (1960). 'Effects of photoperiod and kind of supplemental light on the growth of *Theobroma cacao*', *Proc. 8th Inter-Amer. Cocoa Conf.*, Trinidad, pp. 82–90.

PIRINGER, A. A., DOWNS, R. J. and BORTHWICK, H. A. (1958). 'Effects of photoperiods on *Rauvolfia*', *Am. J. Bot.*, **45**, 323–26.

REES, A. R. (1961). [Report on oil-palm research in Plant physiology Division.] *W. Afr. Inst. Oil Palm Research*, 9th Ann. Rep. 1960–61, pp. 89–94.

REES, A. R. (1964). 'The flowering behaviour of *Clerodendrum incisum* in Southern Nigeria', *J. Ecol.*, **52**, 9–17.

RICHARDS, P. W. (1952). *The Tropical Rain Forest.* Cambridge. University Press.

RICHARDS, P. W. (1969). 'Speciation in the tropical rain forest and the concept of the niche', *J. Linn. Soc. (Biol)*, **1**, 149–53.

RICHARDS, P. W., TANSLEY, A. G. and WATT, A. S. (1939, 1940). 'The recording of structure, life-form and flora of tropical forest communities as a basis for their classification', *Imp. For. Inst., Oxford*, Paper no. 19: 1–19. (Also published in *J. Ecol.*, **28**, 224–39).

RICHARDSON, S. D. (1957). 'Bud dormancy and root growth in *Acer saccharinum*', in: *The Physiology of Forest Trees*, ed. K. V. Thimann, pp. 409–25. New York. Ronald Press.

ROBBINS, R. G. and WYATT-SMITH, J. (1964). 'Dry land forest formations and forest types in the Malayan peninsula', *Malay. Forester*, **27**, 188–216.

RUTTER, A. J. and WHITEHEAD, F. H. (Eds.) (1963). *The Water Relations of Plants*. Oxford. Blackwell.

SACH, J. von (1887). *Lectures on the Physiology of Plants*. Oxford. University Press.

SCHIMPER, A. F. W. (1898). *Pflanzengeographie auf physiologischer Grundlage*. Jena. (English edition *Plant Geography on a Physiological Basis*, 1903, Oxford. Translated by W. R. Fisher, eds. P. Groom and I. B. Balfour.)

SCHIMPER, A. F. W. and FABER, F. C. von (1935). *Pflanzengeographie auf physiologischer Grundlage*. Jena.

SCHMITZ, A. (1963). 'Aperçu sur les groupements végétaux du Katanga', *Bull. Soc. r. Bot. Belg.*, **96**, 233–460.

SCHNELL, R. (1950). *La forêt dense*. Paris. Paul Lechevalier.

SCHNELL, R. (1952). 'Contribution à une étude phytosociologique et phytogéographique de l'Afrique occidentale: les groupements et les unités géobotaniques de la région guinéenne', *Mém. de l'IFAN*, **18**, 41–234.

SCHNELL, R. (1961). 'Le problème des homologies phytogéographiques entre l'Afrique et l'Àmerique tropicales', *Mém. Mus. nat. Hist. nat., Paris*, sér. B, **11** (2), 137–242.

SCHULZ, J. P. (1960). *Ecological Studies on Rain Forest in Northern Suriname*.' Amsterdam. North Holland.

SEQUEIRA, L. and STEEVES, T. A. (1954). 'Auxin inactivation and its relation to leaf-drop caused by the fungus *Omphalia flavida*', *Pl. Physiol. Lancaster*, **29**, 11–16.

SIMON, S. V. (1914). 'Studien über die Periodizität der Lebensprozesse der in dauernd feuchten Tropengebieten heimischen Bäume', *Jb. wiss.Bot.*, **54**, 71–187.

SINGH, L. B. (1960). *The Mango*. London. Leonard Hill.

STAHEL, J. (1971). 'Anatomical investigations on buttresses of *Khaya ivorensis* A. Chev. and *Piptadeniastrum africanum*' (Hook.f.)Brenan, *Holz Roh-u. Werkstoff*, **29**, 314–18.

STEENIS, C. G. G. J. van (1958). 'Basic principles of rain forest sociology', in: *Study of Tropical Vegetation*, pp. 159–65. Paris. UNESCO.

STOCKER, O. (1960). 'Physiological and morphological changes in plants due to water deficiency', in: *Plant-Water Relationships in Arid and Semi-arid Conditions*, pp. 63–104. Paris. UNESCO.

SYMINGTON, C. F. (1933). 'The study of secondary growth on rain forest sites', *Malay. Forester*, **2**, 107–17.

TAYLOR, C. J. (1960). *Synecology and Silviculture in Ghana.* Edinburgh. Thomas Nelson.

THIMANN, K. V. (1962). 'Research on plant physiology in the tropics', *Bull. Ass. trop. Biol.*, **1**, 86–9.

TIXIER, P. (1966). *Flore et végétation orophiles de l'Asie tropicale.* Paris. Soc. d'édition d'enseignement supérieur.

WALTER, H. (1962). *Die Vegetation der Erde in ökologischer Betrachtung.* Bd. 1. *Die tropischen und subtropischen Zonen. Jena.* [English edition '*Ecology of Tropical and Subtropical Vegetation*', 1971. Edinburgh. Oliver & Boyd.]

WALTER, H. and LIETH, H. (1960–67). *Klimadiagramm-Weltatlas.* Jena. G. Fischer.

WAREING, P. F. (1969). 'The control of bud dormancy in seed plants', *Symp. Soc. exp. Biol.*, **23**, 241–62.

WAREING, P. F., HANNEY, C. E. A. and DIGBY, J. (1964). 'The role of endogenous hormones in cambial activity and xylem differentiation', in: *Formation of Wood in Forest Trees*, ed. M. H. Zimmermann, pp. 323–344. New York. Academic Press.

WARMING, E. (1909). *Oecology of Plants.* Oxford. (Translated P. Groom and I. B. Balfour.)

WEBB, L. J. (1958). 'Note on the studies on rain forest vegetation in Australia', in: *Study of Tropical Vegetation*, UNESCO Kandy Symp. Proc., pp. 171–74.

WENT, F. W. (1957). *The Experimental Control of Plant Growth.* Chronica Botanica.

WENT, F. W. (1962). 'Some problems of plant physiology in the tropics', *Bull. Ass. trop. Biol.*, **1**, 90–91.

WHITMORE, T. C. (1962a). 'Studies in systematic bark morphology. I. Bark morphology in the Dipterocarpaceae', *New Phytol.*, **61**, 191–207.

WHITMORE, T. C. (1962b). 'II. General features of bark construction in Dipterocarpaceae', *New Phytol.*, **61**, 208–20.

WHITTAKER, R. H. (1970). *Communities and Ecosystems.* London. Macmillan.

WIGHT, W. and BARUA, D. N. (1955). 'The nature of dormancy in the tea plant', *J. exp. Bot.*, **6**, 1–5.

WILKINSON, G. (1939). 'Root competition and silviculture', *Malay. Forester*, **8**, 11–15.

WYATT-SMITH, J. (1953). 'A note on the vegetation of some islands in the Malacca Straits', *Malay. Forester*, **16**, 191–205.

WYATT-SMITH, J. (1954). 'Storm forest in Kelantan', *Malay. Forester*, **17**, 5–11.

ZAHL, P. A. (1964). 'Malaysia's giant flower- and insect-trapping plants', *National Geographic*, **125**, 681–97.

ZAHNER, R. (1968). 'Water deficits and growth of trees', in: *Water Deficits and Plant Growth*, vol. 2, ed. T. T. Kozlowski, pp. 191–254. New York. Academic Press.

ZIMMERMANN, M. H. and BROWN, C. L. (1971). *Trees: Structure and Function*. Berlin. Springer.

ZIMMERMANN, M. H., WARDROP, A. B. and TOMLINSON, P. B. (1968). 'Tension wood in the aerial roots of *Ficus benjamina* L.', *Wood Sci. and Tech.*, **2**, 95–104.

Note:–

A recent additional source of information on many ecological aspects of tropical forest is:–

ODUM, H. T. and PIGEON, R. F., Eds. (1970). *A tropical rain forest: a study of irradiation and ecology at El Verde, Puerto Rico*. Washington, D.C. Atomic Energy Comm. 3 Vols.

Index of plant species

Numbers in **bold** type refer to illustrations. See also Table 4.4, page 71 for representative genera in the various Tropical-forest regions.

General Index

Numbers in **bold** type refer to illustrations